International Conference on Wind Propulsion 2019

London, United Kingdom
15-16 October 2019

ISBN: 978-1-7138-0872-5

Printed from e-media with permission by:

Curran Associates, Inc.
57 Morehouse Lane
Red Hook, NY 12571

Some format issues inherent in the e-media version may also appear in this print version.

The Royal Institution of Naval Architects
8-9 Northumberland Street
London, WC2N 5DA
United Kingdom

Phone: 020 7235 4622
Fax: 020 7259 5912

publications@rina.org.uk

Additional copies of this publication are available from:

Curran Associates, Inc.
57 Morehouse Lane
Red Hook, NY 12571 USA
Phone: 845-758-0400
Fax: 845-758-2633
Email: curran@proceedings.com
Web: www.proceedings.com

CONTENTS

***Unavailable at time of going to press**

International Conference on Wind Propulsion

15-16 October 2019, London, UK

In association with:

IWSA
International Windship Association

THE ROYAL INSTITUTION OF NAVAL ARCHITECTS
8-9 Northumberland Street, London, WC2N 5DA, Telephone: +44 (0)20 7235 4622

International Conference on Wind Propulsion

15-16 October 2019, London, UK

In association with:

International Windship Association

THE ROYAL INSTITUTION OF NAVAL ARCHITECTS
8-9 Northumberland Street, London, WC2N 5DA, Telephone: +44 (0)20 7235 4622

WIND ASSISTED PROPULSION SYSTEMS AND THE ROLE OF A SHIP CLASSIFICATION SOCIETY

U. Hollenbach, H. Hoffmeister and **J. Wienke**, DNV GL, Germany

SUMMARY

As an independent third-party consultant, DNV GL provides dedicated simulation studies on comparing the performance, reliability and economic impact of different sailing rig concepts in a high-level assessment. In more detailed studies a comparison of the reduction of green-house gases on different shipping routes throughout the year can be assessed.

Each newbuilding must fulfil mandatory EEDI limit values. What is almost unknown in the shipbuilding community is that wind propulsion systems can become a significant measure to fulfil future stricter limits. EEDI rules and regulations offer the possibility considering the effect of WAPS (Wind Assisted Propulsions Systems) in the EEDI calculation. The assumptions in the EEDI rule set regarding a global wind matrix with reference to the global shipping routes are explained and discussed.

DNV GL has traditionally been dealing with certification and engineering of sailing rigs and uses these synergies for offering certification and engineering services within the wind propulsion segment. Recently technical standards for wind assisted propulsion systems and a new additional class notation for the ships carrying such devices has been developed. These services and standards are presented in the third part of the paper.

NOMENCLATURE

[Symbol]	[Definition] [(Unit)]
dof	Degrees of freedom
DTMB	David Taylor Model Basin
ECMWF	European Centre for Medium-Range Weather Forecasts
EEDI	Energy Efficiency Design Index
HSVA	Hamburg Ship Model Basin
IMO	International Maritime Organization
ITTC	International Towing Tank Conference
MARPOL	International Convention for the Prevention of Marine Pollution from Ships
MEPC	Marine Environment Protection Committee
SEEMP	Ship Energy Efficiency Management Plan
WAPS	Wind Assisted Propulsion System

1. INTRODUCTION

DNV GL traditionally has been offering certification services around sailing rig technology for more than a century. Former board member of Germanischer Lloyd, Ernst Middendorf, wrote a technical book about the design and construction of Tallship Rigs. After over 100 years, it still stands valid and has been the basis for the DNV GL Standard "Tallship Rigs" which is enjoying worldwide reception in that industry.

Since over 25 years ago, DNV GL, former Germanischer Lloyd, has been accompanying the pioneering ages when carbon masts, and later carbon rigging, were introduced in the super yacht industry, by having developed technical benchmark standards for the purpose of impartial assessment and certification. Today, many flag state authorities refer to these certification programs, when yachts are used in commercial purpose.

German Naval Architect Peter Schenzle realized a sailing cargo ship in the mid 1980´s, featuring an "Indosail" rig. It demonstrably featured fuel savings in the 70% range.

Even before, in the late 1960´s, Wilhelm Prölss developed the Dynarig for cargo ships, which 45 years later saw the light on the mega yacht "Maltese Falcon", a modern interpretation of a "square rigger", but being able to be handled by a small crew by a drastic simplification and automation in comparison to the classical square rigs. Recent IMO regulation require attention for greener shipping. One of the options reducing the emission of green-house gases and fulfilling ever stricter requirements are to use the wind as source of supplementary energy.

Figure 1: Wilhelm Prölss with a model of the Dynarig

In the book "Windschiffe" [1] from 1988, the German authors prophesied: "...Using the wind means using the most environmentally friendly energy source. Saving (fossil) fuel means emitting fewer pollutants. Time will come, in which this aspect will be considered more valuable than pure commercial interest."

In his article "Wind Ships in the 21st Century?" [2] from 2010 Peter Schenzle summarizes current concepts for wind propulsion of ships. He argues that sea transportation particularly lends itself for first steps away from carbon combustion, for three reasons:

- Its uniquely low energy demand could be largely covered from solar sources;
- Wind is easily available at sea and can directly drive ships without transformation losses;
- Weather routeing and energy management can largely compensate for the variable input.

Figure 2: Modern square rigger "Maltese Falcon"

Finally, this new legislation issued by IMO is asking for just this. Cutting greenhouse gas emissions, providing mandatory targets to reach has driven ship owners and operators to re-think their propulsion systems. One of the possible solutions to cope with the requirements is to install a wind assisted propulsion system on board to relieve the main combustion engine and to save fuel and thus reduce emissions.

Among ship owners there is an increasing interest about the chances and possibilities of wind assisted propulsion.

2. CONCEPT ASSESSMENT

DNV GL provides a third-party assessment on the performance, reliability and economic impact of sailing rig concepts.

In the following the procedure assessing different wind propulsion systems is described in principal.

2.1 ROUTES

In high level studies dedicated routes are not considered. Different systems are compared using an average wind speed and evenly distributed encounter angles.

In detailed assessment studies several routes are considered in the simulations. This can be one-way connections or return connections or combinations of both, which are extracted from the Marine Traffic website. If desired, alternative routes can be analyzed, which are longer but likely offer more promising weather conditions for sailing, e.g. Tradewinds versus direct route going Transatlantic westbound.

The effect of route optimization has not yet considered in such studies, but this could be included as well.

2.2 WEATHER DATA

Each route is segmented into pieces with a maximum length of 60 nautical miles. For each segment the wind speed and direction as well as the significant height of combined wind waves and swell are determined.

Nowadays historic weather data is available through various providers on the internet. The European Centre for Medium-Range Weather Forecasts (ECMWF) Copernicus portal [3] offers access to global weather data from 1979 to present. The database results from the ERA5 fifth generation reanalysis based on observations across the world and resolves hourly at a 1° x 1° grid for mean ocean waves. Included quantities are significant height of combined wind waves and swell, mean wave direction, mean wave period and u- and v-components of the wind in 10 m reference height.

Reference can be obtained from here: https://confluence.ecmwf.int/display/CKB/ERA5+data+documentation#ERA5datadocumentation-HowtociteERA5

Historic data for the u- and v-component of surface currents can be obtained from the CFSv2 database, which is accessible through the Research Data Archive at the National Center for Atmospheric Research, Computational and Information Systems Laboratory. The data is available from 2011 to present and the reanalysis resolves 6-hour averages for the entire globe at a 0.5° x 0.5° grid.

Reference: Saha, S., et al. 2011, updated daily. NCEP Climate Forecast System Version 2 (CFSv2) 6-hourly Products. Research Data Archive at the National Center for Atmospheric Research, Computational and Information Systems Laboratory. https://doi.org/10.5065/D61C1TXF

2.3 VOYAGES

Each voyage along a certain route is simulated starting on the first of every month over a period of ten years.

This implies 120 runs (12 months x 10 years) for each route. The weather data obtained from the databases refers to the time and the predicted position, corresponding to the targeted sailing speed of the ship. With up to one hundred and twenty segments for each route and considering different target sailing speeds the number of simulations can easily reach more than 100,000 simulations to be performed. This is done on a computer cluster with approximately 6,000 CPUs.

2.4 POWER PREDICTION

The power prediction is computed by the velocity prediction program "FS-Equilibrium". With it the equilibrium for 6 degrees of freedom (dof) can be determined. However, for such studies 5 dof are regarded sufficient, as the trim is assumed to be constant.

2.4 (a) Hydrostatics

The ship model is simulated at one draft, which relates to a specified displacement. The geometry model includes appendages like a skeg and rudder(s).

2.4 (b) Hydrodynamics

The frictional resistance of the hull is computed for each condition and depends on the wetted surface. The surface roughness and the correct correlation with the full-scale ship are ensured by applying a reasonable correlation allowance (Ca).

The appraisal of the residual resistance is based on a regression model for similar hulls.

The transversal force and hence the related additional drag are computed with the approach by Schenzle (HSVA 1989) for sail ships without long fin keels.

Buoyancy and resistance of the rudder and skeg are computed by the approach by Whicker and Fehlner (DTMB Report 933, 1958) for hydrofoils with small aspect ratios.

The propeller is modelled as a Wageningen B series propeller.

2.4 (c) Aerodynamics

The drag and lift coefficients of the most important WAPS (Wind Assisted Propulsion Systems) as "traditional square rig", "modern square rig" (Dynarig), "rotor sails" (Flettner), "dynamic kite" (Skysails) or "soft/rigid wing sails" in principal publicly available.

For example, the drag and lift coefficients of the "modern square rig" (Dynarig) of the "Maltese Falcon" are available by publication by Dijkstra [4], performance data of a state-of-the-art "rotor ship" (Flettner) of the "E-Ship 1" by Schmidt [5] and Vahs [6].

The drag and lift coefficients are tuned in accordance to the aspect ratios of the masts and in accordance to the expected efficiency of the rig concept (e.g. number of masts, etc.). The aerodynamic drag of the masts and spreaders is computed by the transversal and longitudinal projected areas and the corresponding drag coefficients.

The aerodynamic drag coefficient of the hull is adopted from wind channel tests for similar ships and used with the transversal and longitudinal projected areas of the projected vessel.

For each aerodynamic force component, the apparent wind speed is used at the height of the area center. To compute the wind speed at various heights, the wind profile is adjusted to match the wind speed at 10 m.

2.4 (d) Added resistance in a seaway

The added resistance in a seaway is based on a regression model and uses the main parameters of the vessel and the significant wave height at the predicted position.

For this analysis the added resistance is assumed to be independent from the main wave direction.

2.4 (e) Simulation behaviour

If the ship reaches the target sailing speed (typically 8 knots, 9 knots or 10 knots) in the direction of the next waypoint then the vessel is purely sailing.

Tacking against the wind is not part of this study.

If the targeted sailing speed cannot be reached though, then the engine supports the cruise until the target speed is reached. If the WAPS does not produce a positive thrust while motor sailing, then the vessel is assumed running without WAPS and only by motor at the target speed.

If necessary, the sail area is reduced not to exceed a limiting heeling angle.

3. ENERGY EFFICIENCY CERTIFICATION

The assessment of a WAPS with regard to the EEDI certification is completely different, mainly since this index should allow comparability with reference lines and thus with similar ships as well. This means that a route specific assessment as presented in the preceding chapter is not suitable, but the available power of all WAPS is related to a global wind probability matrix.

The EEDI was introduced by IMO in 2013 as a measure to reduce the GHG emission [MARPOL Annex VI, Chapter 4] [7]. The index gives the CO_2 emission from main and auxiliary engines divided by the transport work of the ship which is the product of the ship's capacity and the ship speed. The CO_2 emission of the engines is calculated by the product of installed power, the engine's specific fuel oil consumption and the fuel specific conversion factor giving the relation between consumed mass of fuel and emitted mass of CO_2 [Regulation 20]. A detailed description of the calculation method of the attained EEDI including all correction factors for special ship types and design conditions is given in resolution MEPC.308(73) [8].

3.1 REFERENCE LINES AND REQUIRED EEDI

Based on averaging over the existing fleet, reference lines were determined for different ship types. These reference lines give the required EEDI as function of the capacity. Four different phases were defined, starting with phase 0 in 2013. For ships ordered from beginning of 2015, phase 1 became relevant and the following phases 2 and 3 start in intervals of 5 years later respectively. For each phase, reduction factors were defined which are in general 10% related to the base line, so that the reduction factor is 10% for phase 1, 20% for phase 2 and 30% for phase 3. Vessels ordered today must fulfil the required EEDI of phase 1, whereas from beginning of next year phase 2 is valid [Regulation 21].

The EEDI applies to almost all new ships with a capacity larger than 400 gross tonnage. Exception are several ship types with non-conventional propulsion and cargo ships with ice breaking capability [Regulation 19]. Non-conventional propulsion includes diesel-electric propulsion, turbine propulsion and hybrid propulsion systems.

The easiest way to fulfil stricter limit values for the EEDI is to reduce the installed engine power. When a smaller main engine is installed, the transport work will decrease with the lower ship speed; but since the relation between ship speed and propulsion power is approximately a function of power to three, the CO_2 emission will be reduced stronger and the EEDI value will improve.

The potential of reducing the main engine power is limited as a certain available power is required for safe navigation in adverse weather conditions. In MEPC.232(65) minimum propulsion power is defined for bulk carrier, tankers and combination carrier [9].

Existing designs can be improved by installation of energy saving devices, such as ducts, pre-swirl devices, fins or propeller boss cap fins. Modification of propeller and rudder often offers potential for improvements as well. All these improvements will be regarded in the EEDI calculation by a higher ship speed for the same installed engine power and improve the EEDI value accordingly.

For new designs the efficiency can be improved by a line optimisation. This improvement is regarded in the same way as the modifications described above, meaning that a higher speed can be achieved for the same power, thus the transport work is improved and the EEDI value is lower.

3.2 INNOVATIVE TECHNOLOGIES

With introduction of phase 2 but with phase 3 at the latest, these methods will not be sufficient anymore for several ship types. The way to comply with required EEDI in future is to apply so-called innovative technologies. WAPS are one of these innovative technologies.

Different categories of innovative energy efficiency technologies are distinguished [MEPC.1/Circ. 815] [10]. On the one hand there are systems to reduce the main engine power, on the other hand there are systems to reduce auxiliary power. The systems that effect the main engine power are divided into category A and B. Category A includes technologies that cannot be separated from overall performance of the vessel, e.g. low friction coating or the already mentioned improved designs of propeller or rudder. These systems are considered in the EEDI formula by the higher speed for the same installed engine power as described above.

Category B covers systems that can be treated separately from the overall performance of the vessel, e.g. WAPS or hull air lubrication systems. To include the effectiveness of these systems into the EEDI calculation an additional term is inserted in the nominator. The power provided by the category B system multiplied by the specific fuel oil consumption of the main engine and the conversion factor of the fuel is subtracted to get the emission reduction related to the innovative technology. Thus, the total CO_2 emission is reduced but the installed engine power and the according ship speed are unchanged. The result is a lower EEDI value.

Category B systems are divided into systems that are always available, e.g. the lubrication system can be switched on all the time, and systems that are only temporarily available as the WAPS depending on the wind occurrence. This availability is regarded by an availability factor of each innovative energy efficiency technology which is below 1 for WAPS.

Innovative technologies can also be applied on board to reduce the required auxiliary power by generating electricity. These systems belong to category C. Again, a differentiation is made between systems that are always available as for example waste heat recovery systems and systems that are only temporarily available as photovoltaic cells. A system specific availability factor must be determined. In the same way as for category B systems the provided power is multiplied by the specific fuel oil consumption but of the auxiliary engines this time and the according fuel conversion factor. The resulting CO_2 emission is deducted from the CO_2 emission of the auxiliary engines and hence the attained EEDI value is reduced. On cargo ships the auxiliary power is typically 5% or less of the installed main engine power. So, the saving potential for category B systems is distinctly larger than for category C systems.

For example, the application of wind turbines on board ships to generate electric power is regarded as innovative technology. As the first passenger ferry in the world, the Stena Jutlandica, was equipped with two wind turbines at the prow in 2011. These turbines contribute to a reduced fuel consumption and hence CO_2 emission by supplying electric energy (category C) and by impacting (supposed reduction) the air resistance of the deckhouse (category A).

3.3 WIND ASSISTED PROPULSION SYSTEMS

The reduction of the CO_2 emission of the main engine is calculated from the available effective power of the installed WAPS. This power is determined from a system specific and ship speed dependent force matrix and the global wind probability matrix. The global wind probability matrix is defined in MEPC 62//INF.34 [11]. It is determined from the wind data of 106 locations along the main global shipping routes as shown in figure 3. For each wind speed up to 25 m/s (10 Bft) in steps of 1 m/s each and for each angle relative to the ship bow, covering 360° in steps of 5°, the probability is defined. Thus, the matrix contains 26*72=1872 elements.

It is important to note, that the angle to which the global wind probability matrix refers is already the relative wind angle. It was determined from combination of the ship heading at the 106 locations and the true wind direction measured there.

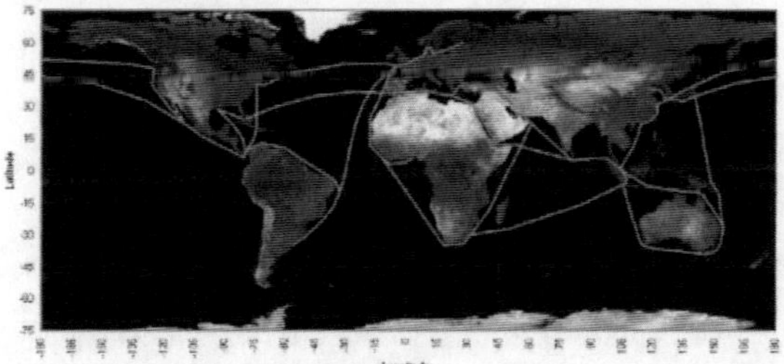

Figure 3: Main ship trading routes to be considered in EEDI calculation [Source: MEPC 62/INF.34]

The force matrix of the WAPS has the same dimension as the global wind probability matrix, meaning 1872 elements. Each element can be determined from the force coefficient in longitudinal ship direction C_X, the sail area A, the apparent wind speed V_A and the air density ρ as follows:

$$F_X = \frac{\rho}{2} \cdot C_X \cdot A \cdot V_A^2$$

The force coefficient C_X can be determined from lift coefficient C_L and drag coefficient C_D by considering the apparent wind angle ε:

$$C_X = C_L \cdot \sin\varepsilon - C_D \cdot \cos\varepsilon$$

Depending on the apparent wind speed the force matrix becomes a function of the ship speed at the same time. The force coefficients are typically determined by wind tunnel model tests. For Flettner rotors also stationary full-scale tests are performed instead.

In the EEDI formula not only this part describing the power delivered by the WAPS is considered, but there is also an according term to determine the power demand of the WAPS deducted to get the effective power.

3.4 VERIFICATION OF THE EEDI

DNV GL acts as verifier of the EEDI. The verifier is an organization duly authorized by an Administration to conduct the surveys and certification of the EEDI. The certification of the EEDI is divided into two separate steps:

1. Preliminary assessment of the EEDI during design phase of the ship (before cutting of the first steel plate). Aim of this assessment is to ensure that the design can fulfil the required EEDI (preferably with a sufficient margin to allow for slightly worse performance of the real ship than predicted).

2. Final assessment of the EEDI after performance assessment during speed trials. Only after conducting the speed trials of the ship and applying appropriate corrections for ambient conditions, the speed power curve of the ship is known for the trial draught. Results of the model tank test will be used to conclude from the sea trials results to the EEDI reference draft (maximum summer load draught for all ship types except for container ship where the reference draught is the 70% deadweight draught).

The Technical File that must be submitted for the preliminary EEDI verification should include an outline of the WAPS and the resulting EEDI value by considering the WAPS. A detailed calculation process of the force matrix and results of the according performance tests should be included. WAPS as category B systems can be treated separately from the overall performance of the vessel. Thus, the impact of the WAPS on the EEDI can already be verified during the preliminary EEDI verification.

For the final EEDI, which is often urgent since it must be verified between sea trials and delivery of the vessel, only an adjustment of the applied force matrix to the updated ship speed is required. It will be confirmed as well that the configuration of the installed WAPS is the same as applied in the preliminary verification but a witness of the operation of the WAPS on board the ship is not required for the EEDI verification.

3.5 OUTLOOK

The application of the global wind probability matrix implies that no route specific assessment is possible and that a differentiation for the different WAPS is restricted. This causes that for some cases the power reduction in the EEDI can be significantly lower than the real power savings in operation when applying all the potentials described in preceding chapter 2.

During MEPC 74 in May 2019, a discussion about how to consider the impact of WAPS started based on an example given by China. A VLCC was equipped with a hard airfoil sail. The EEDI decreased by 1.6% by considering the global wind probability matrix. But when applying a wind probability matrix reduced to the middle east to far east route on which the vessel operates, the EEDI decreases by 8.4%.

It should be kept in mind that the EEDI is a design value. Any operating measures like selecting only suitable sailing routes and the use of weather routing optimization is not yet considered. Regarding future regulations it is discussed to give further incentives introducing innovative technologies. Thus, it might be that the quite conservative approach to consider WAPS will be more realistic in the future and will result in larger reductions of the considered power when WAPS are installed.

4. CERTIFICATION / CLASSIFICATION

DNV GL felt obliged to respond on the many inquiries for independent assessment of those systems and developed a Standard for the certification of WAPS and an additional class notation, called "WAPS" [12].

The DNV GL Standard ST-0511 "Wind assisted Propulsion Systems" [13] will be published later this year but is available as a draft and can already be used for the purpose of certification. It has been developed typically for "Flettner" type rotors and associates and for "wing rigs" and associates. The technical requirements in that Standard are generic enough so that it could be applied for other types of WAPS, in close association.

The core of this Standard is the definition of structural design loads typical for large sail installations on top of the deck of cargo ships. It is distinguished between "in operation" and "out of operation" modes. The following loads are being considered:

- Wind loads (regular and extreme)
- Inertia loads (self-weight effect from ship motions)
- Green sea and spray water loads (for exposed structures)
- Snow and ice loads (in certain areas)
- Precession forces (for rotating systems)

The interaction of the WAPS with the ship´s hull infrastructure can be quite complex and thus load combinations of the above listed loads need to be assessed.

Evidence is requested in this standard, how WAPS cope with particularly extreme wind conditions, combined with other loads, particularly considering the large areas exposed to wind and possibly also green water on deck.

This is the reason why in before-hand, particularly for novel concepts, a risk analysis must be carried out, addressing all aspects of design, equipment and operation in order to identify detrimental effects in early design phase. By the standard, it is recommended to perform this risk analysis using a FMEA (Failure Mode Effect Analysis), giving rise to the possible consequences.

Depending on the "dominance" of the WAPS onboard a ship, i.e. the relative size between WAPS and ship, amongst the structural failure of the system, intact stability for the ship can get critical, when excessive aerodynamic loads can´t be controlled.

While technical aspects like:

- structural design
- intact stability
- machinery installations
- electrical installations
- control systems

are subject to DNV GL certification /classification, other aspects which are also required for the ship operation are being addressed:

- line of sight from bridge
- radar blind sector
- manoeuvrability

are part of the flag authority's approval program.

DNV GL Standard ST-0511 is intended for the certification of WAPS. It obviously primarily serves the WAPS Designers / Manufacturers for the purpose of having their systems independently assessed and certified. The Certification is offered in different levels and depths, depending on the demand:

a) "AIP"-Approval in Principle:

   An approval in principle is an independent assessment of a concept within an agreed framework, led by the Society, by confirming that the design is feasible, and no significant obstacles exist to prevent the concept from being realized.

   The basic design and operational concept will be reviewed and approved. The functionality of the system and restrictions of operation will be assessed and approved.

   The Society may, on a case-by-case basis, require to perform an AiP for novel designs before engaging in type approval design or other certification activities.

b) "TADC"- Type Approval Design Certificate:

The Type Approval Design Certificate is a compliance document validating the compliance with all design requirements in all technical disciplines as defined by the standard.

c) "TAC"- Type Approval:

A Type Approval Certificate is a compliance document including the scope of the TADC but amended with DNV GL production assessment and survey requirements on materials, components and the WAPS, once completed.

One of the above certifications may suit a WAPS manufacturer demonstrating compliance with a recognized standard to the desired level.

If a ship owner decides to apply for the (new) additional class notation "WAPS", the minimum certification level required for the WAPS system is the TADC. Additional to that requirement, technical safety aspects of the ship´s periphery associated with and affected by the installation of a WAPS unit will need to be approved according to the relevant class rules. These are the technical installations integrated into the ships systems, such as electrical and automation, machinery and the foundation as well as safety aspects such as ship stability.

5. CONCLUSIONS

The authors explain the role of the hip classification society DNV GL in the concept assessment, in the energy efficiency certification and in the certification and classification of Wind Assisted Propulsion Systems (WAPS).

6. REFERENCES

1. BERTHOLDT, J., RIESCH, H., 'Windschiffe', *VEB Verlag Technik*, 1988.

2. SCHENZLE, P., 'Wind Ships in the 21st Century? – Current concepts for wind propulsion of ships', *STG Sprechtag "Innovative Schiffe" in Kiel*, March 2010.

3. European Centre for Medium-Range Weather Forecasts (ECMWF) Copernicus Portal: https://cds.climate.copernicus.eu/cdsapp#!/dataset/reanalysis-era5-single-levels?tab=overview

4. DIJKSTRA, G., 'The Maltese Falcon: the realisation', *The International HISWA Symposium on Yacht Design and Yacht Construction*, 2004.

5. SCHMIDT, A., 'E-Ship 1 – A Wind-Hybrid Commercial Cargo Ship', *4th International Conference on Ship Efficiency, Hamburg*, September 2013.

6. VAHS, M., 'Flettnerrotor senkt Brennstoffkosten', *Schiff & Hafen, Nr. 2*, 2019.

7. IMO, MARPOL Annex VI, Chapter 4, 'Regulations on energy efficiency for ships', 2018.

8. IMO, resolution MEPC.308(73), '2018 Guidelines on the method of calculation of the attained energy efficiency design index (EEDI) for new ships', 2018.

9. IMO, resolution MEPC.232(65), '2013 Interim guidelines for determing minimum propulsion power to maintain the manoeuvrability of ships in adverse conditions', 2013.

10. IMO, MEPC.1/Circ. 815, '2013 Guidance on treatment of innovative energy efficiency technologies for calculation and verification of the attained EEDI', 2013.

11. IMO, MEPC 62//INF.34, 'Global wind specification along the main global shipping routes to be applied in the EEDI calculation of wind propulsion systems', 2011.

12. DNV GL Rules for Classification, Part 6, Chapter 2, Section 12, Wind Assisted Propulsion Systems WAPS, ed. 7-2019.

13. DNVGL-ST-0511, Wind Assisted Propulsion Systems (draft standard).

7. AUTHORS BIOGRAPHY

U. Hollenbach holds the current position of a Principal Engineer in the Fluid Engineering team at DNV GL. He has been working in different positions in the shipbuilding industry. His previous experience includes basic design of ships, weight estimation, power predictions, ship theory calculations, intact- and damage stability calculations, hull form design and optimization, perform and evaluation of model tests, preparation, conduct and evaluation of sea trials, member of ISO working group, member of ITTC specialist committee.

H. Hoffmeister holds the current position of a Senior Principal Engineer at DNV GL. Since 1993, he has been involved in the assessment of marine structures, yachts and rigs, developed several standards and guidelines, amongst which are the GL Guidelines for Structural Design of Racing Yachts and those for Superyacht Carbon Rigs. His particular field of expertise is the evaluation of composite structures. Parallel to his employment at Germanischer Lloyd, he was member of the design team of "United Internet Team Germany" for the Americas Cup 2007 as the team's rig designer.

J. Wienke holds the current position of a Principal Engineer at DNV GL. He performs measurements during sea trials of newbuilding ships and has attended more than 200 sea trials world-wide during the last 17 years. The main item of his experience is speed trials and he joined the according ISO working group. He is involved in the EEDI verification by approval work, surveys and training of surveyors. Previous experience includes work in a wave channel and analysis of slamming events.

PREDICTED FUEL-SAVINGS FOR A FLETTNER ROTOR ASSISTED TANKER USING COMPUTATIONAL FLUID DYNAMICS

L. Jones and **M. Prince,** Wolfson Unit M.T.I.A., UK
D. Hudson, University of Southampton, UK
J. Cocks, Shell Shipping and Maritime, UK

SUMMARY

This paper presents a case study of the effectiveness of Flettner rotors as a wind-assist system for tanker hull forms with particular focus on the aerodynamic modelling of the devices. It summarises a series of computation fluid dynamics simulations carried out to evaluate the impact of parametric changes in Flettner rotor design, including the impact of multiple rotors operating in combination, and the influence of the ship induced flow field. The results of these simulations have been applied within a velocity performance prediction approach to estimate reductions in thrust required from the propeller, and hence changes in engine power and fuel consumption in realistic operations. The analysis highlights Flettner rotors as a means of providing significant levels of power assistance dependent upon true wind speed, heading and ship speed.

NOMENCLATURE

C	Ratio between the diameter of the rotor end-plate and the diameter of the main rotor body
C_D	Drag coefficient
C_L	Lift coefficient
C_M	Moment or torque coefficient
D	Rotor diameter (m)
E	Rotor efficiency
L	Rotor lift (N)
P_r	Power required to rotate a rotor (W)
P_t	Thrust power produced by a rotor (W)
U	Free-stream velocity (m/s)
VPP	Velocity Prediction Program
VR	Velocity Ratio: Circumference velocity/wind speed (-)

1. INTRODUCTION

The recent increase in the use of Flettner rotors on commercial vessels has been supported by an increase in the activity of researchers in aerodynamic numerical and experimental modelling, eg. [1] and [2].

The objective of this research was to extend numerical modelling and assessment tools to aid performance predictions in terms of wind assisted powering of ships with Flettner rotors, with some approaches being transferrable to the assessment of other shipborne wind assistance systems. To this end, the performance of full scale Flettner rotors has been investigated using computational fluid dynamics (CFD) for a variety of wind conditions, geometries and placement scenarios, with the intention of demonstrating factors that may affect rotor performance in real-world installations. Results from the CFD analysis were then incorporated into a performance prediction model to assess potential power savings, taking into account factors such as the aero/hydrodynamic resistance of the vessel and expected wind conditions.

2. COMPUTATIONAL FLUID DYNAMICS METHODOLOGY

CFD simulations were conducted on the IRIDIS 5 supercomputer at the University of Southampton using the OpenFOAM toolbox, with details as follows:

- Simulations were conducted using an unsteady Reynolds averaged Navier-Stokes (URANS) based transient solver.
- The rotor was modelled as rotating by specifying a moving-wall boundary condition.
- Simulations were conducted at 1:1 scale, hence at full-scale Reynolds numbers (up to 5×10^6).
- Simulations were conducted with a minimum domain size of 80 rotor diameters wide by 100 rotor diameters long by 24 rotor diameters high.
- The air was assumed to be 15 degrees Celsius.
- Turbulence effects were accounted for using a kappa-omega SST turbulence model, with the viscous sublayer modelled using wall functions.

The simulations are comparatively expensive in computational terms due to the nature of the problem. The flow around the rotor is an unsteady bluff body flow, and hence the required temporal integration for the simulations scales with the number of flows past the cylinder (i.e. D/U). Conversely, because the surface velocity of the cylinder is typically multiples of the freestream velocity, the mesh resolution near the wall must be fine and the time step small relative to the temporal integration length. Ultimately, these conflicting requirements mean that the simulations require considerably more computational expense than flow past, for example, a static cylinder, wing or conventional sailing yacht rig. Therefore, whilst a case could be made for using large or direct eddy simulation (LES or DES) due to the bluff-body flow physics present, this is precluded for reasons of pragmatism, since these methods would be prohibitively computationally expensive for the parametric analysis conducted here.

Each simulation was run for at least 24 hours using 32 processors, and a total of 104 simulations have been conducted, resulting in 80,000 processor hours in total. Simulations typically used nominally 20 million cells per rotor.

3. ANALYSIS OF BASELINE ROTOR GEOMETRY

A 'baseline' rotor geometry is defined as possessing diameter 5m and height 30m, with a rotating end-cap (or so-called 'Thom disk') possessing twice the rotor diameter (i.e. C=2). This rotor geometry is intended to represent the scale appropriate to large vessels, e.g. of the order of 200m LWL. Unless otherwise specified, for example when investigating the effect of geometry changes, all simulations use the baseline rotor geometry. The effect of wind speed and rotation rate is investigated by simulating the baseline Flettner rotor geometry at a range of realistic wind speeds and rotation rates. The rotation rate was varied between a minimum of 50RPM and a maximum of 236RPM, however all results are presented in terms of the velocity ratio, *VR*, defined as the tangential velocity at the rotor surface divided by the free-stream velocity. The inflow condition was specified as a uniform velocity distribution, and a visual example is provided in Figure 1. Plotting the results in dimensional form (Figure 2), illustrates that as the wind speed increases the forces acting upon the rotor, as well as the torque required to drive the rotor, increase significantly. Plotting these results in coefficient form collapses them to nominally a single curve each (Figure 3 & 4), broadly independent of wind speed and dimensional rotation rate. This means that results can be scaled to predict rotor performance at other wind speed and velocity ratios, for example when conducting route analysis.

The L/D ratio of the rotor reaches a maximum at a low velocity ratio in the range, $1<VR<2$, and then decreases with increasing velocity ratio, however L/D is not an especially illuminating quantity in terms of quantifying rotor performance for ship propulsion. In this context, the rotor efficiency, defined as the thrust power produced by the rotor divided by the rotational power required to drive it, is perhaps the most important metric. Under normal operating conditions a Flettner rotor will produce more power than required to drive it, hence this parameter may also be thought of as a power multiplication factor, and the terms are used interchangeably here. When referring to rotor efficiency, it should be noted that thrust power is a function of apparent wind speed and direction, and hence can only be determined in the context of a rotor on a vessel travelling at a known speed, with known wind conditions. For the purposes of assessing rotor performance in isolation, a pseudo power multiplication factor can be defined as $E=LU/P_r$, with L the rotor lift and U the freestream velocity. This quantity is a maximum at the smallest velocity ratio tested, and then decreases with increasing velocity ratio (figure 5). For the baseline rotor configuration, E=1 occurs at a velocity ratio of 10. This means that below a velocity ratio of 10, the rotor will always provide more thrust energy than it consumes in order to rotate it. However, when considering this result it should be borne in mind that the flow conditions investigated via CFD are similar to a rotor in optimum conditions – e.g. a beam reach, whereas in practice a proportion of the lift will be delivered as sideforce instead of thrust. The precise amount will vary depending upon wind angle, however the effect will be to reduce the cut-off point (in terms of velocity ratio) at which the rotor becomes uneconomical.

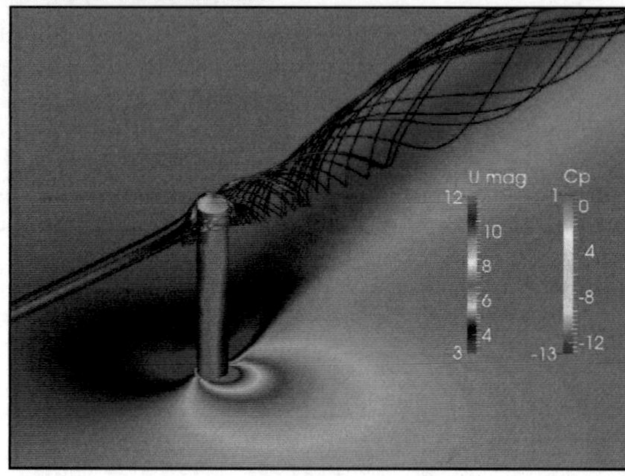

Figure 1: Illustration of a Flettner rotor with no end-cap in a uniform flow. The rotor is coloured by pressure coefficient, the near-wall plane is coloured by velocity magnitude, and streamlines are shown illustrating the vortex originating at the rotor tip

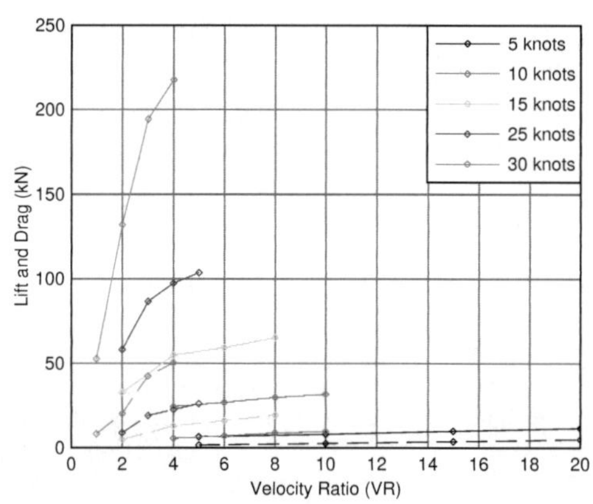

Figure 2: Variation of lift (solid) and drag (dashed) with velocity ratio for the baseline rotor

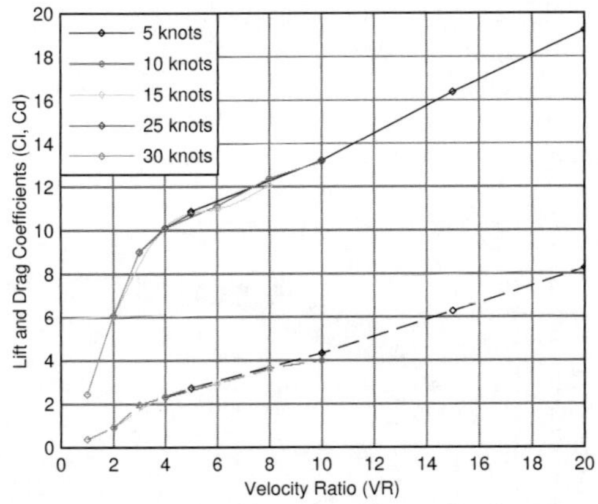

Figure 3: Variation of lift coefficient (solid) and drag coefficient (dashed) with velocity ratio for the baseline rotor

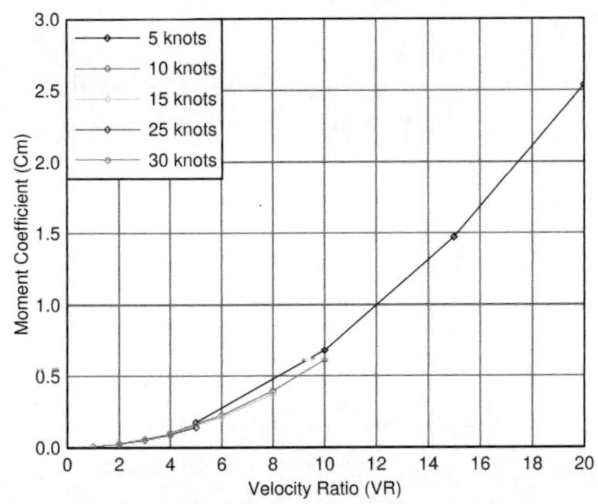

Figure 4: Variation of moment coefficient with velocity ratio for the baseline rotor

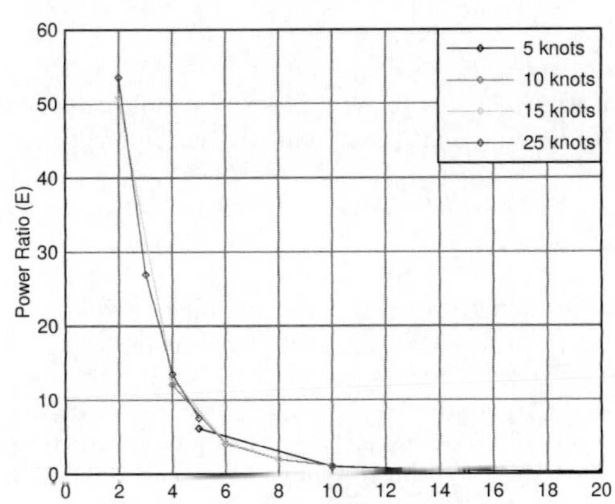

Figure 5: Variation of power multiplication factor with velocity ratio for the baseline rotor

4. INFLUENCE OF ASPECT RATIO

The influence of aspect ratio upon rotor performance is investigated by simulating a series of Flettner rotors possessing different heights and diameters, all subject to a uniform 15 knot wind profile at velocity ratios of 2, 4, 6 and 8, and possessing an end-cap diameter ratio of 2.

The effect of increasing rotor aspect ratio is to decrease the drag coefficient significantly, increase the lift coefficient slightly (Figure 6), and to decrease the torque coefficient (Figure 7). The net effect is that the power multiplication factor increases with increasing aspect ratio (Figure 8) meaning that, ignoring all other factors and constraints, the ideal shape for a Flettner rotor is as large an aspect ratio (i.e. as tall and slender) as possible. It should be noted however that the benefit of increasing aspect ratio decreases as the aspect ratio increases – the rotor performance asymptotes to the 'ideal' situation of an infinite aspect ratio.

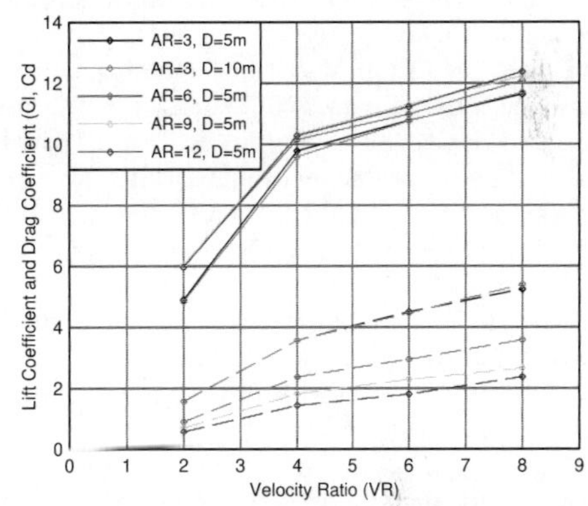

Figure 6: Variation of lift (solid) and drag (dashed) coefficients with velocity ratio for rotors posessing differing dimensions

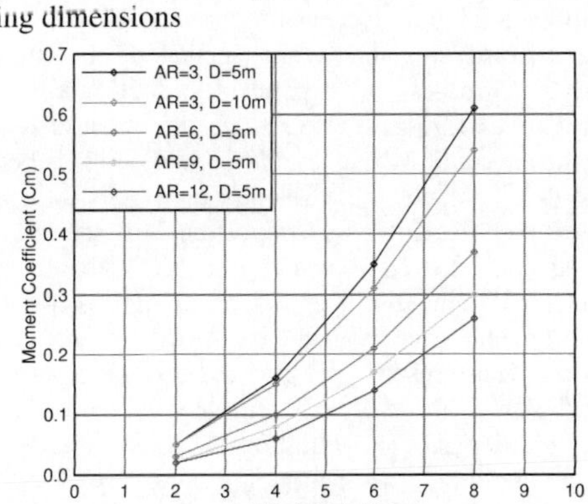

Figure 7: Variation of moment coefficient with velocity ratio for rotors posessing differing dimensions

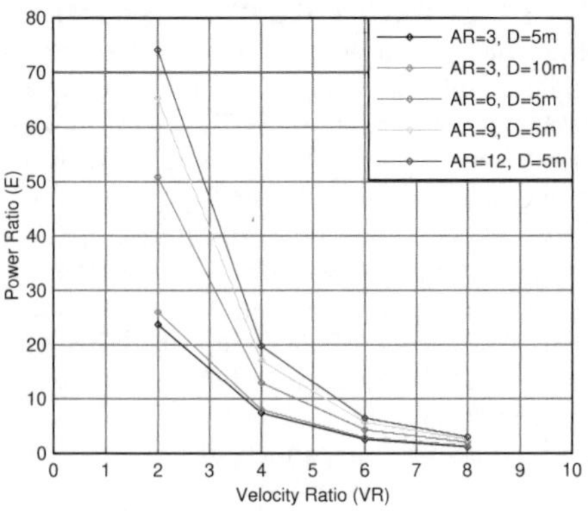

Figure 8: Variation of power multiplication factor with velocity ratio for rotors posessing differing dimensions

5. INFLUENCE OF ROTOR END PLATE

Earlier studies on Flettner rotors report that the addition of an end plate (or 'Thom disk') to the end of a rotor can decrease the rotor drag and will increase the rotor lift [3]. This effect is investigated for a full scale rotor by varying the relative size of the end-plate whilst keeping the thickness constant. The end plate diameter ratio is increased from C=1 (i.e. no end-plate) to C=4, for which simulations were conducted at a single windspeed of 15 knots and velocity ratio of 6. It can be seen in Figure 9 that as the end-plate diameter increases from C=1 to C=2 the drag coefficient increases to a maximum before dropping with further increases in plate diameter, whereas the lift coefficient exhibits a step-change increase between C=1.5 and C=1.75, and then a more gradual increase with further increases in end plate diameter. Perhaps most importantly, the moment coefficient increases significantly with increasing end plate diameter, and so the power multiplication factor decreases monotonically (Figure 10). These results imply that, whilst increasing the end-plate diameter increases lift significantly, particularly when moving to C=1.75-2, it also reduces the rotor efficiency, even for the smallest disc tested (C=1.25). This is because as the disc diameter increases, not only does the surface area of the disc increase proportional to the diameter squared, but the edge velocity of the disc increases proportional to the disc diameter, resulting in a compound increase in skin friction and hence torque required to drive the rotor. These results appear to suggest that a rotor with no end-plate is the most efficient, although not the most powerful, design. The question therefore arises as to whether holding the end-plate stationary (i.e. non-rotating), whilst operating the rotor will lead to an increase in rotor efficiency. In order to investigate this, simulations of the baseline rotor geometry were conducted with a static end-plate (C=2), and are plotted in Figure 10.

The effect of holding the end-plate stationary is to reduce the lift and drag coefficients slightly, but to significantly reduce the moment coefficient. The net effect is a large increase in the power multiplication factor, nominally by a factor of two at C=2, as noted by [4]. In fact, the effect of holding the end-plate stationary is to increase the power ratio over and above that observed for the rotor with no end-plate (*E*=9.1 as compared to *E*=8.1 at *VR*=6). These results represent the highest power multiplication factors observed for all end-plate geometries tested. The CFD analysis therefore predicts a clear performance benefit of holding the end-plate stationary, however the drawback to this approach would be the increased system complexity required to hold the end-plate static.

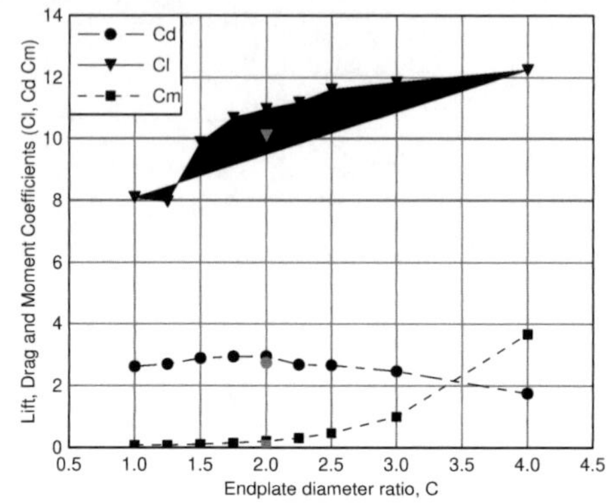

Figure 9: Variation of lift, drag and moment coefficients with endplate diameter ratio for rotating (black) and static (red) end plates

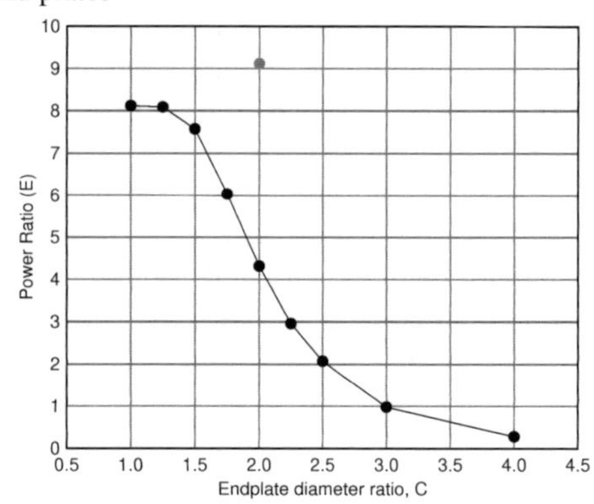

Figure 10: Variation of power ratio with endplate diameter ratio for rotating (black) and static (red) end plates

6. INTERFERENCE EFFECTS OF MULTIPLE ROTORS

The influence of rotor separation is investigated by conducting simulations of a pair of Flettner rotors and varying the separation distance and separation vector. Simulations have been conducted translating the rotors both in the streamwise (x) and lateral (y) directions, with the rotor axes separated by 2.5, 3, 4 and 6 rotor diameters (i.e. 12.5, 15, 20 and 30m). All simulations were conducted with a free-stream velocity of 15 knots and nominal velocity ratio of 6.

Figures 11 and 12 show that for all cases the mean drag coefficient is higher than that of a single rotor and the mean lift coefficient is lower than for a single rotor.. Accordingly, the mean power multiplication factor is also lower than for a single rotor (Figure 13). This means that the net effect of rotor interference is to reduce system performance. The behaviour of the rotors differs depending on whether the rotor separation is in the streamwise or lateral directions, however.

When the rotor separation is in the streamwise direction (Figure 15), the drag acting on the rotors is the most significantly affected quantity, with the downstream rotor exhibiting significantly increased drag, and the upstream rotor significantly decreased drag. Inspecting contours of pressure at the midplane suggests that this is because the location of minimum pressure (i.e. suction) moves slightly toward the downstream side of the downstream rotor, and to the upstream side of the upstream rotor, as compared to the case of a rotor in isolation. Lift is also affected (reduced), but less significantly. The overall system performance appears to be particularly sensitive to streamwise separation, because even at the largest separation tested (30m centre-to-centre) the mean power multiplication ratio had not recovered to that of a single rotor, being 10% lower, and the mean drag was still 50% higher than for a single rotor.

When the rotor separation is in the lateral direction (Figure 16) it is the lift produced by the rotors that is affected most significantly, with the rotor on the high speed/suction side exhibiting dramatically increased lift, but the rotor on the low speed/pressure side exhibiting dramatically decreased lift. This appears to be because the superposition of rotor flows increases the maximum velocity in the vicinity of the rotor on the high speed side, but decreases it on the low speed side (compare Figures 14 to 16). The mean lift, however, is close to that of a single rotor. The performance of laterally spaced rotors appears less sensitive than for streamwise spacing, because by 30m spacing the mean power multiplication factor had returned to that of a single rotor, although the mean drag coefficient was still 30% higher. This is in keeping with [2] identifying that drag coefficient is sensitive to changes in pressure distribution. These results show that rotor performance can be strongly affected by interference effects, particularly for rotors aligned in the streamwise direction. It is clear that rotors should ideally be placed as far away from one another as possible, and it is conceivable that there may be merit in considering locating rotors in, for example, a staggered pattern.

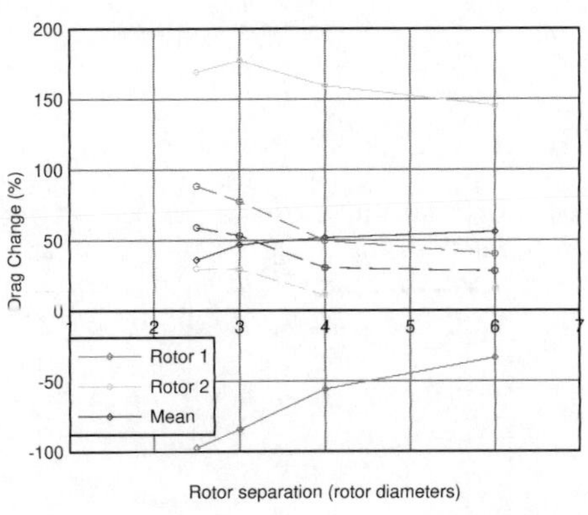

Figure 11: Drag change as a function of longitudinal (solid) and lateral (dashed) rotor separation, relative to a rotor in isolation

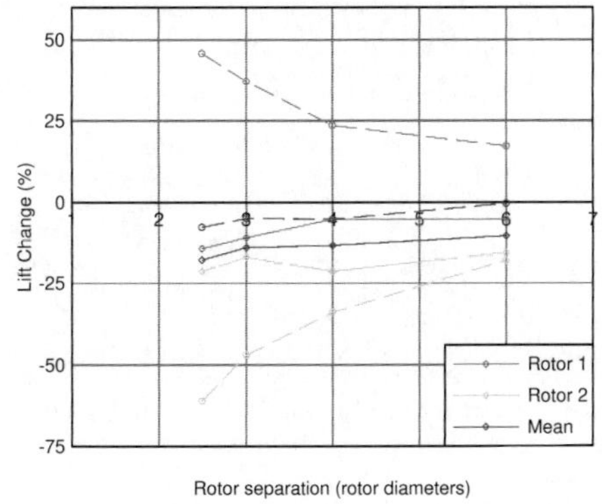

Figure 12: Lift change as a function of longitudinal (solid) and lateral (dashed) rotor separation, relative to a rotor in isolation

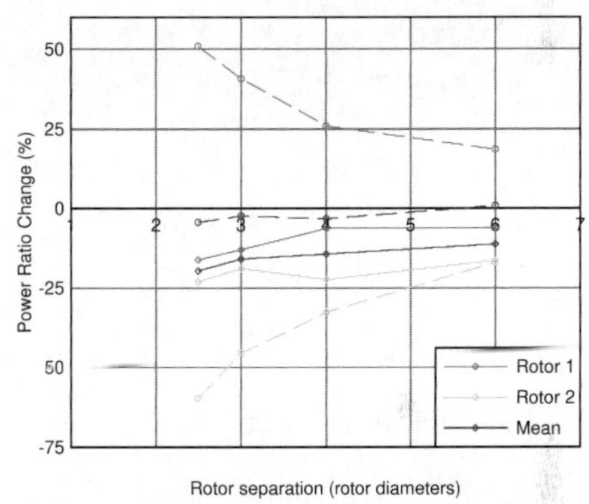

Figure 13: Power ratio change as a function of longitudinal (solid) and lateral (dashed) rotor separation, relative to a rotor in isolation

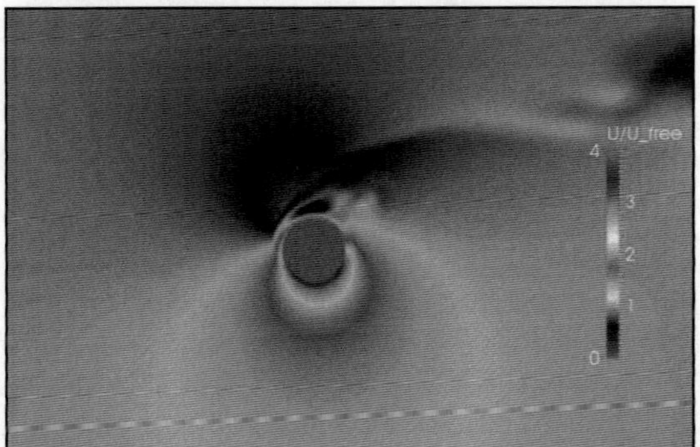

Figure 14: Velocity ratio plotted at the mid-height for a single rotor at U=15 knots and VR=6

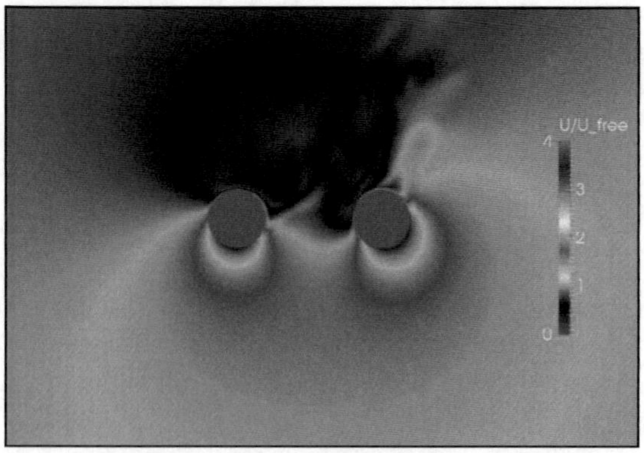

Figure 15: Velocity ratio plotted at the mid-height for two rotors with a streamwise separation of 7.5m, at U=15 knots and VR=6

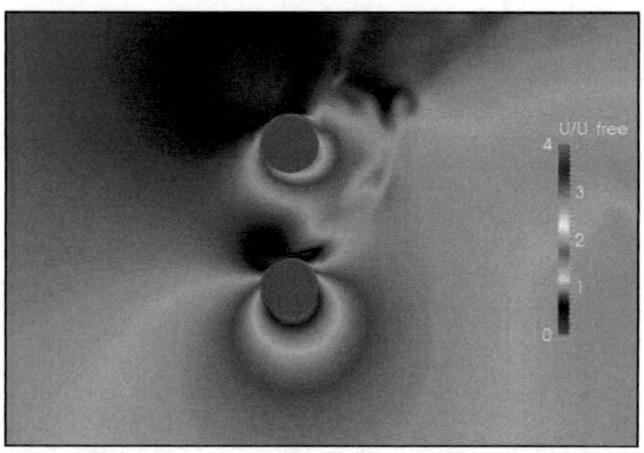

Figure 16: Velocity ratio plotted at the mid-height for two rotors with a lateral separation of 7.5m, at U=15 knots and VR=6

7. INFLUENCE OF SUPERSTRUCTURE PROXIMITY

The influence of nearby superstructure is investigated by conducting simulations of a rotor in proximity to a cuboid obstacle, i.e. a box, representing a generic superstructure element causing a flow blockage. The box possesses width 30m, depth 10m and height 10m, and is placed with the long side facing the rotor. A three dimensional view of the configuration is provided in Figure 17. The rotor and box were simulated for a range of separation distances and blockage angles around the rotor. The angles are defined such that 0 degrees is immediately upstream of the rotor, 90 degrees is beside the rotor, on the high speed/suction side, and 270 degrees is beside the rotor on the low speed/pressure side. All simulations were conducted with a free-stream velocity of 15 knots and velocity ratio of 6.

When the separation gap is small, i.e. one rotor diameter (or 5m), the rotor lift does not ever exceed that of a rotor in isolation irrespective of obstacle orientation relative to the rotor. Therefore the presence of a nearby obstacle acts only to degrade rotor performance (Figure 18). The rotor efficiency performance is most severely degraded when the obstacle is directly upstream of the rotor (0 deg.), and at +/-45 deg. upstream either side of the rotor centreline. The performance is also degraded significantly, but to a lesser extent, when the obstacle lies downstream of the rotor. These results suggest that, if it is desired to optimise the performance of a vessel for reaching conditions, it is preferable for superstructure to be located fore or aft of a Flettner rotor as opposed to on the port or starboard side.

The effect of varying the distance of the box from the rotor was investigated for lateral placements only (i.e. +/- 90 degrees). The results (Figure 19) suggest that placing an obstacle on the pressure side of the rotor at close proximity is more harmful than placing an obstacle on the suction side of a rotor. However, the performance detriment decreases rapidly as the box is moved further away, and in fact an efficiency increase of nominally 5% is reported at a separation gap of three rotor diameters (15m).

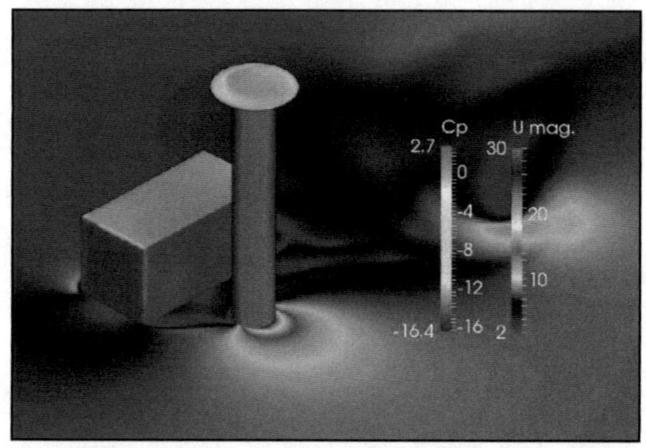

Figure 17: Image of a Flettner rotor with box located on the pressure/low velocity side. The rotor is coloured by the pressure coefficient, whilst the near-wall plane shows the velocity magnitude. The air is flowing from the bottom-left to the top-right of the image.

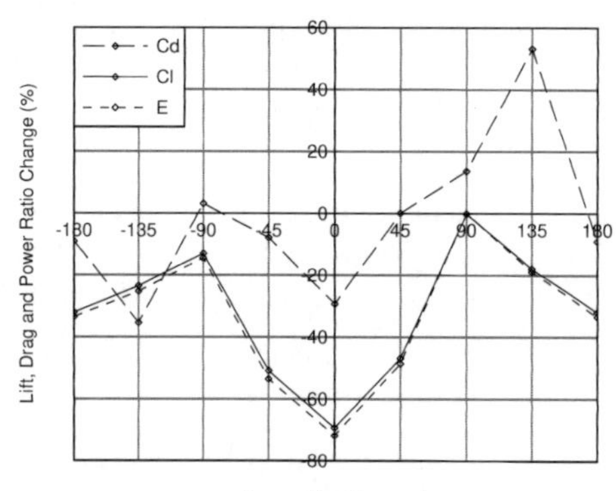

Figure 18: Drag, lift and efficiency change as a function of box orientation, for a separation gap of one rotor diameters, relative to a rotor in isolation

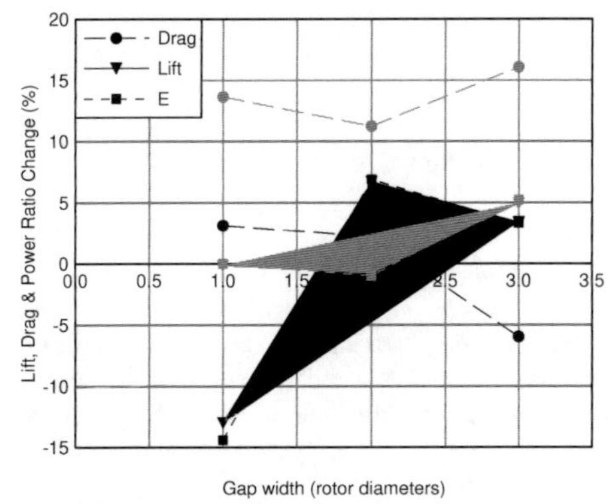

Figure 19: Drag, lift and efficiency change as a function of a separation gap width for a box on the pressure side of the rotor (black) and on the suction side of the rotor (red), relative to a rotor in isolation

8. INFLUENCE OF SITING UPON A SHIP HULL

The influence of locating a Flettner rotor on a ship form, as opposed to on a flat surface, is investigated by simulating a single rotor on a simple ship geometry, without superstructure, at a variety of wind angles, at a wind speed of 15 knots and velocity ratio of 6. The detail of the ship form is not critical, it is rather the blockage effect of the vessel on the wind flow that is important. To this end, a generic hullform was created, possessing 200m LOA, 40m beam and 10m freeboard. The rotor was located nominally 45m from the bow and 15m from the centreline of the vessel, and the complete configuration is illustrated in Figure 20. The wind angles are defined such that 0 degrees corresponds to flow from bow to stern, and 90 degrees corresponds to flow from port to starboard.

The effect of including the ship structure is to introduce a secondary flow pattern which interacts with the flow around the Flettner rotor. Effectively the ship creates a blockage (modified flow field) which, particularly for wind angles of +/-90 degrees, creates a large region of flow separation, comprising a slower moving region of air over the deck and downstream of the vessel (Figure 21) and encompassing the Flettner rotor. Where the wind angle is 0 or 180 degrees the ship imparts a significantly smaller blockage to the flow, and the Flettner rotor does not sit within a low speed/separated region of flow (Figure 22).

The net effect of the ship presence is to reduce the rotor efficiency at all points of sail, as seen from Figure 23. The rotor efficiency is lowest at +/-90 degrees, and -135 degrees, which are angles at which the ship presents a large frontal area to the flow, and creates the largest wake over the deck. At these angles since the rotor lies in a region of decelerated flow, the effective velocity ratio of the rotor increases, causing the efficiency of the rotor to decrease in line with the trends identified in section 2. When the ship is orientated at 0 or 180 degrees the rotor does not lie in a large separated region and the rotor efficiency is correspondingly higher, however since the Magnus effect produces thrust perpendicular to the wind direction, the Flettner rotor would not produce useable thrust under these conditions. These results indicate that in practice the flow around a rotor installed on a ship may differ significantly from simple uniform or power law boundary layers. In particular, reaching (true wind angles of the order of 90°) conditions may lead to the flow over the rotor being decelerated, which may potentially require the use of a lower rotation rate in order to maintain a velocity ratio corresponding to efficient rotor operation.

Figure 20: View of a simple ship hull with Flettner rotor, with wind angle of -45 degrees. The rotor and ship hull are coloured by pressure coefficient, whilst the near-wall plane is coloured by velocity magnitude.

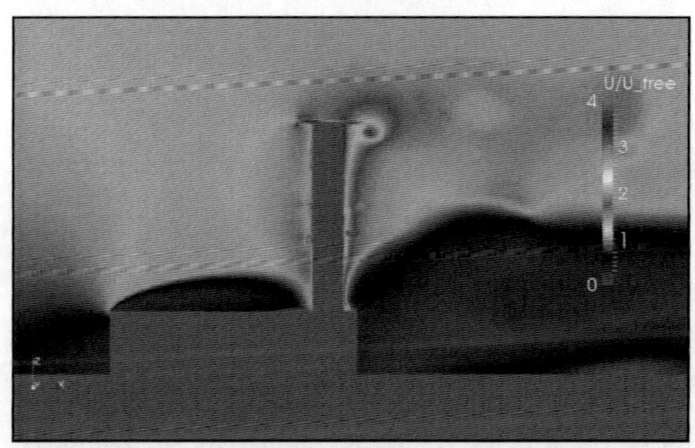

Figure 21: Non-dimensional velocity plotted on a plane through the centre of a rotor located on a simple ship, for a flow angle of 90 degrees at U=15knots and VR=6. The air is flowing from left to right.

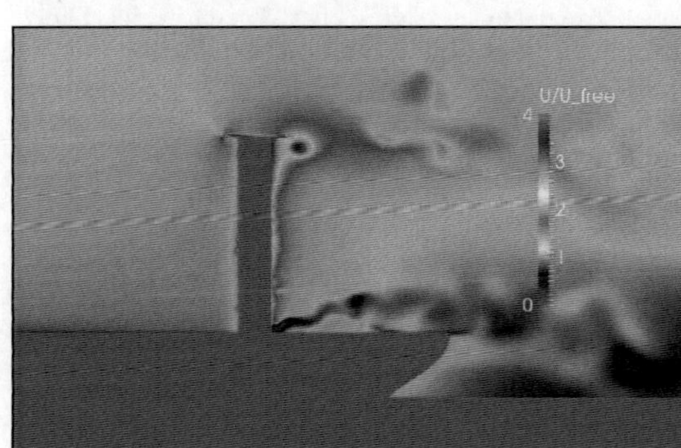

Figure 22: Non-dimensional velocity plotted on a plane through the centre of a rotor located on a simple ship, for a flow angle of 180 degrees at U=15knots and VR=6. The air is flowing from left to right.

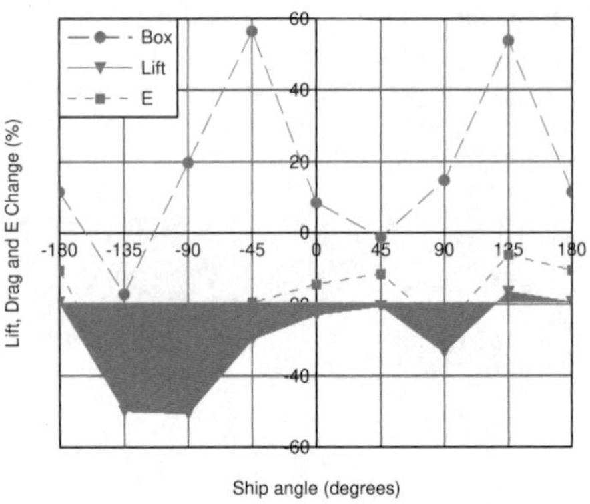

Figure 23: Drag, lift and efficiency change as a function of wind direction for a rotor mounted on a generic ship hullform, relative to a rotor in isolation.

9. COMMENTS ON OPTIMAL ROTOR SPECIFICATIONS

When considering a Flettner rotor installation on a vessel one of the most critical questions regards what height and diameter should the rotors should be. It has been demonstrated that rotors with high aspect ratios are more efficient, however in practice the maximum height of a rotor installation is likely to be limited by the requirement to navigate underneath bridges. Assuming that the rotor height is fixed by operational or structural considerations, the next dimension to specify is the rotor diameter. Increasing the rotor diameter for a fixed rotor height has a number of effects. For example, as diameter is increased the efficiency of the rotor decreases but the absolute lift produced by the rotor increases. Clearly a very efficient rotor is not practical if it cannot produce a meaningful amount of lift, and likewise a powerful rotor is not practical if it is inefficient, but between these extremes there is a viable range of rotor diameter. Furthermore, as the rotor diameter increases the drag of the rotor also increases, meaning that during time periods when wind conditions are not favourable the rotor will represent a bigger detriment to vessel efficiency. Therefore, it is likely that the greater percentage of time a vessel is expected to experience favourable wind conditions, the larger the optimum rotor diameter may be. It is clear then, that the optimum rotor dimensions and placement are likely to depend upon the size and speed of the vessel, as well as the particular route(s) navigated, and hence wind conditions experienced, by the vessel. So in practice, whilst the analysis of Flettner rotors in isolation can illustrate behavioural trends, any optimisation of parameters such as rotor diameter or spatial placement cannot be achieved by considering the performance of rotors in isolation. In order to optimise the various aspects of a rotor installation a system-wide approach must be undertaken, considering both the hydrodynamic and aerodynamic performance of both the vessel and rotor installation, as well as the drivetrain performance and, critically, the route and expected wind conditions experienced by the vessel.

10. CASE STUDY

In order to assess the impact of differences in rotor performance, two types of generic ship form are evaluated; a LNG carrier and a LR2 tanker. Principal dimensions of these vessels are provided in Table 1.

A number of configurations were taken from the CFD analysis described previously and a selection of them are detailed in Table 2.

		LNG	LR2
LOA	m	276	245
Bmax	m	46	42
T	m	11.4	15.475
Displacement	tonnes	105000	125000
Freeboard	m	14.5	14

Table 1: Principal vessel parameters

Configuration	Units	A	B	C	D	E	F
Ship Type		LNG	LNG	LNG	LNG	LR2	LR2
Aspect Ratio		6	9	4.17	6	6	6
No. of rotors		4	4	4	4	4	2
Rotor Height	m	30	36.67	25	25	30	30
Rotor Diameter	m	5	4.08	6	4.17	5	5
Rotor Limit	rpm	180	221	150	216	180	180
Single Ref. Area	m^2	150	150	150	104.2	150	150
Total Ref. Area	m^2	600	600	600	416.7	600	300

Table 2: Modelled rotor configuration details

11. PERFORMANCE PREDICTION

In order to assess the aerodynamic performance impact resulting from CFD generated data a velocity performance prediction (VPP) approach was used. A simplified assessment approach was selected yielding results in terms of power saving, i.e. % reduction in delivered engine power. It is noted that other authors have tackled some of these aspects in more detail, as can be seen in ref [1].

An existing sailing yacht based VPP, Wolfson Unit 'WinDesign' was modified and used for this case study. This amalgamates force and moment contributions from the hydrodynamic and aerodynamic components and solves for the ship speed and associated stability for a range of true wind speeds, headings, engine and rotor contributions. The basic sailing aspects of this are in keeping with [5].

11.1 HYDRODYNAMIC FORCE MODELLING

The hydrodynamic force model main components were as follows:

$$R_T = R_U + R_H + R_I - T_P$$

- R_T: total hydrodynamic resistance
- R_U: upright resistance; based on the power prediction for single screw merchant ships from the Wolfson Unit Powering Program (calm water predictions).
- R_H: resistance due to heel based on a regression of in-house model tank data of ship forms in a 'sailing' condition.

- R_I: induced drag based on a regression of in-house model tank data of ship forms in a 'sailing' condition.
- Tp: Thrust from drivetrain contribution

These were amalgamated into a hydrodynamic force model that was input into the VPP.

11.2 AERODYNAMIC FORCE MODELLING

The aerodynamic force model was created based on the previously presented CFD derived data. The majority of the simulations were carried out with a rotor in isolation and only a limited number with the rotor(s) located on the deck of the ship. The reductions in lift and increase in drag highlighted in sections 6 & 8 have been used to create factors to modify isolated rotor data to incorporate the ship induced and multiple rotor influences.

11.3 SOLUTION PHASE

This solution phase seeks the equilibrium between the aerodynamic and hydrodynamic longitudinal forces, transverse forces and roll moments at a range of true wind speeds and angles, engine thrust and rotor VR settings.

11.4 ASSUMPTIONS

In order to conduct this analysis a number of assumptions have been made, these include:

- A constant full load displacement condition was used for all simulations
- A rotational velocity limit of 180 rpm has been applied to 5m diameter rotors with proportional limits applied to other rotor diameters. All rotor data and limits are summarised in Table 2.
- Powering of the rotor(s) has been accounted for in the power saving calculations and efficiency factors have been applied in the conversion of effective to delivered power.
- Reduction of rotor performance due to siting on a ship geometry has been included, including the loss of lift and increase in drag highlighted in section 8.
- Performance coefficients used assume that the positioning of the rotor(s) will avoid significant adverse interactions.
- An aerodynamic windage model (including interactions with superstructure) of the rotors when not in use (i.e. when motoring upwind) has been included.
- Location of the rotor(s) and use of the vessel's hydrodynamic lift generating appendages to maintain a steady course has been ignored. It is acknowledged that it is vital to have yaw balance between the hydrodynamic and aerodynamic induced moments for the vessel to be able to sail.
- The impact of ship motions on the aerodynamic and hydrodynamic behaviour of the rotors have been ignored

11.5 RESULTS

The VPP analysis resulted in predictions of percentage power savings for specific constant ship speeds at a range of true wind speeds and angles. An example is presented in Table 4.

	True Wind Speed (knots)					
True Wind Angle (deg)	5	10	15	20	25	30
0	-2.6	-5.0	-8.1	-11.3	-14.3	-16.9
10	-2.5	-4.9	-7.9	-11.1	-14.1	-16.7
20	-2.4	-4.7	-7.5	-10.5	-13.4	-16.0
30	-2.3	-4.3	-6.8	-9.6	-12.3	-14.8
40	-2.1	-3.8	-5.9	-8.3	-10.7	-13.1
50	-1.9	-3.2	-4.8	-6.7	-8.7	-10.8
52	-1.8	-3.1	-4.6	-6.4	-8.3	-10.3
60	-1.6	-2.6	-3.7	-5.0	-6.5	-8.1
70	-1.4	-2.0	-2.6	-3.4	1.8	8.0
80	-1.2	-1.4	-1.4	9.0	26.8	31.4
90	-1.0	-1.0	1.4	19.5	39.6	43.0
100	-0.8	-0.6	4.9	27.9	47.6	65.4
110	-0.6	-0.4	6.5	31.1	51.1	70.0
120	-0.5	-0.2	6.1	30.7	53.6	71.0
135	-0.3	-0.1	3.0	25.6	48.7	66.6
150	-0.3	0.0	0.2	15.5	39.3	59.4
160	-0.2	0.0	0.0	8.0	30.4	50.4
170	-0.2	0.0	0.0	1.5	20.2	39.5

Table 3: Breakdown of percentage power savings for configuration A at constant 12 knots ship speed

11.6 ROUTE INFORMATION

Weather data used in this study is based on the summation of data recorded from a number of LNG carriers in operation over a 2 year period. This is representative of actual ship activities as it includes combinations of various routes, the true wind angle and speed distribution and is summarised in Table 4.

When combined with the VPP analysis it results in the power savings per configuration and steady ship speed ranges as presented in Table 5. These highlight the reduction in rotor propulsion assistance with increasing ship speed resulting from the decrease in apparent wind angle and associated reduction in thrust contribution, with no notable returns at 16 knots of ship speed or above.

The higher aspect ratio rotors, option B, provided a slight benefit (0.26%) in overall power reduction over the datum aspect ratio option (A).

There will be specific routes which would be more beneficial to the operation of rotors, those with predominantly strong winds (i.e. 20 knots and above) from 90° true wind or greater, reflected in the bottom right area of Table 3. This highlights the benefits of Flettner rotors as promising devices in practice for the immediate reduction in Greenhouse gas emissions from shipping.

		True wind speed (knots)					
		0-5	5-10	10-15	15-20	20-25	25-30
True Wind Angle (deg)	0-30	3.2	4.0	3.8	2.6	0.9	0.5
	30-60	2.8	4.0	4.5	3.3	1.4	0.5
	60-90	1.3	3.6	5.3	4.2	2.7	1.0
	90-120	1.1	2.7	4.7	3.8	2.3	1.2
	120-150	1.6	3.0	3.6	4.2	2.3	1.4
	150-180	1.3	2.2	3.0	3.7	1.9	1.1

		True wind speed (knots)					
		30-35	35-40	40-45	45-50	50-55	55-60
True Wind Angle (deg)	0-30	0.2	0.1	0.0	0.0	0.0	0.0
	30-60	0.3	0.1	0.0	0.0	0.0	0.0
	60-90	0.4	0.1	0.1	0.0	0.0	0.0
	90-120	0.4	0.2	0.1	0.0	0.0	0.0
	120-150	1.0	0.6	0.3	0.0	0.0	0.0
	150-180	0.6	0.4	0.2	0.1	0.0	0.0

Table 4: Summary of applied route data

Ship Speed	A	B	C	D	E	F
kts	LNG	LNG	LNG	LNG	LR2	LR2
8	-	-	-	-	12.60%	6.26%
10	8.34%	8.60%	7.89%	5.44%	8.29%	3.17%
12	4.68%	5.01%	4.42%	2.60%	4.48%	1.07%
14	2.03%	2.32%	1.94%	0.77%	1.81%	-0.09%
16	0.36%	0.52%	0.39%	-0.21%	0.21%	-0.57%
18	-0.67%	-0.62%	-0.51%	-0.65%	-0.68%	-0.66%
20	-1.03%	-1.08%	-0.86%	-0.71%	-0.96%	-0.57%

Table 5: Percentage power saving for a range of configurations

12. CONCLUSIONS

The behaviour of Flettner rotors has been investigated by conducting a series of CFD simulations.

It was determined that for a given wind speed, as the velocity ratio is increased the rotor lift, drag and torque all increase, however the rotor efficiency decreased. For the operating conditions simulated the results could be successfully collapsed to a single curve by plotting in non-dimensional coefficient form.

Rotor efficiency was found to increase with aspect ratio, however the optimum rotor dimensions for real-world installations must be determined by a holistic performance analysis, and are likely to be limited by operational constraints.

Adding a rotating end-plate to the rotor was found to increase the lift coefficient but decrease the rotor efficiency. Simulation results suggest that by holding the end-plate stationary, the efficiency can be improved above that of a rotor with no end-plate.

Where rotors are located in close proximity, the effect of the rotor interaction is to decrease the mean rotor performance. This effect is more pronounced where the rotors are orientated in the streamwise direction, where a power multiplication factor penalty of 10% was observed even at a separation of 6 rotor diameters.

Where a rotor lies in close proximity to a flow obstruction such as a box, the net effect is to decrease rotor performance. The performance penalty is largest where the obstacle is upstream of the rotor, and least when the obstacle lies either side of the rotor.

Where a rotor is located on a simple ship geometry, the flow experienced by the rotor can be significantly different to that of an ideal atmospheric boundary layer. In particular, where the free-stream flow lies at/close to 90 degrees the ship hull creates a large decelerated region of flow that may encompass the rotor. Decelerating the flow over the rotor increases the effective velocity ratio, and decreases the rotor efficiency.

The CFD results have been utilised within a VPP analysis process, resulting in predicted power savings of the order of 8% for a 4-rotor vessel at 10 knots based on actual averaged weather route information from vessel operations. The rotors are predicted to be ineffective as a propulsion assistance mechanism at 16 knots of ship speed or greater with these weather data applied.

13. ACKNOWLEDGEMENTS

The authors would like to thank Shell Shipping and Maritime for funding this project and giving permission for it to be published.

14. REFERENCES

1. van der Kolk, N.J., Bordogna, G., Mason, J.C., Desprairies, P., Vrijdag, A. 'Case study: Wind-assisted ship propulsion performance prediction, routing, and economic modelling', *RINA Conf. Power & Propulsion Alternatives for Ships*, 2019.
2. Bordogna, G., Muggiasca, S., Giappino, S., Belloli, M., Keuning, J.A., Huijmans, R.H.M., 'Experiments on a Flettner rotor at critical and supercritical Reynolds numbers', *Journal of Wind Engineering & Industrial Aerodynamics*, 2019.
3. Thom. A., 'On the effects of discs on the air forces on a rotating cylinder', *Reports and memoranda, Aeronautical Research Committee report 1623*, 1935.
4. Badalamenti. C, 'On the application of rotating cylinders to micro air vehicles', *Doctoral Thesis City University*, 2010.
5. Oliver J.C., Claughton, A.R., 'Development of a multi-functional velocity prediction program (VPP) for sailing yachts', *RINA Intl Conf CADAP '95*, 1995.

PERFORMANCE PREDICTION OF WIND PROPULSION SYSTEMS USING 3D CFD AND ROUTE SIMULATION

A. Persson, D.-Q. Li, F. Olsson and **S. Werner**, SSPA Sweden AB, Sweden
U. Dhomé, KTH Royal Institute of Technology, Sweden

SUMMARY

Accurate performance prediction is necessary when designing/optimising wind propulsion systems (WPS). An independent, trustworthy prediction of the energy-saving potential is also needed to support the ship owner's decision to invest in new technology. By using weather statistics along with a mathematical model of ship performance, route simulations can estimate the time and power required for transit of a route. Such simulations are commonly used today to optimise the design and operation of conventional ships. The introduction of WPS poses additional challenges for route simulations. WPS performance must be predicted at all points along the route, with wind of differing velocity and direction. The apparent wind will vary vertically (twist), due to the interaction between the ship velocity and the atmospheric boundary layer. Also, many proposed concepts use multiple WPS, introducing additional complexity, such as independent spin ratios/ sheeting angles. 3D CFD simulations capture the complex physics, including vortex formation and interaction effects, providing accurate performance prediction and an understanding of the flow. However, 3D CFD is costly, and it would not be possible to simulate all conditions at a reasonable cost. We present simplified approaches to modelling of WPS, using a limited number of CFD simulations, either in 2D or 3D, which are then extrapolated such that 3D effects are represented, and all conditions covered. The methodology is demonstrated on rotor sails and wing sails.

NOMENCLATURE

α	Angle of attack (deg)
Γ	Circulation (m²/s)
C_L *and* C_D	Global lift and drag coefficient (-)
C_l *and* C_d	Section lift and drag coefficient (-)
C_x *and* C_y	Section thrust and side force coeff. (-)
D *and* R	Rotor diameter and radius (m)
c	Wing chord length (m)
H	Wing/rotor height (m)
AR	Wing/rotor aspect ratio (-)
SR	Rotor spin ratio, $SR=\omega RU_\infty^{-1}$ (-)
F_{Rx} *and* F_{Ry}	Rotor thrust and side force (N)
V_S	Ship speed (kn)
TWA *and* TWS	True Wind Angle/Speed (°, m/s)
AWA *and* AWS	Apparent Wind Angle/Speed (°, m/s)

1. INTRODUCTION

With the International Maritime Organization's (IMO) target of a reduction of greenhouse gas emissions from the global shipping, the application of a wind propulsion system (WPS) as a retrofit on existing or new-built merchant ships is a promising solution. The introduction of a new WPS technology is, however, a considerable investment. The ship owners often request decision support based on reliable computational predictions of the expected fuel-saving, carried out by a supplier-independent party. Assessment of the aerodynamic performance is also required for the optimisation of WPS design and control system, as well as for addressing structural loads, risks and safety issues.

The performance of a WPS depend on the wind conditions, and therefore the energy efficiency needs to be predicted for all weather conditions that the ship will encounter. This can be accomplished using route simulations that combine the statistical weather along the route with a numerical model of ship performance. This requires prediction of the WPS aerodynamic performance at a large number of conditions, covering all wind velocities and directions along the route. Moreover, the twist of the apparent wind due to the atmospheric boundary layer is unique for every combination of ship speed and true wind. In addition, WPS settings (rotor spin ratio/wing sheeting angle) add to the complexity.

A Computational Fluid Dynamics (CFD) simulation in 3D captures the complex physics, including vortex formation and interaction effects, providing accurate performance prediction and understanding of the flow. However, such computations are costly (up to 3000 CPU hours per condition), and it would not be possible to simulate all conditions at a reasonable cost. However, for most WPS technologies including rotors and wing sails, 2D simulations are not sufficient. The flow is highly three-dimensional with a large interaction effect between the hull and the WPS and between the WPS devices.

To achieve a reliable prediction of the energy-saving at a feasible time and cost, a suitable combination of 3D CFD, 2D CFD and potential flow theory is suggested. The methodology is demonstrated on two commonly investigated WPS technologies; Rotor sails (Flettner rotor) and wing sails. Even though the interaction between multiple devices and the ship hull can be considerable, this study uses single WPS devices without the presence of a ship hull. In this way, the pure effect from the simplifications can be studied separately.

2. TEST CASES

2.1 SINGLE ELEMENT WING SAIL

A wing sail is a rigid wing-shaped structure placed vertically on the ship. The most well-known application is on America's Cup racing catamarans, but it has also been tested on cargo vessels. The single element test case studied here is based on the Wallenius Marine Sailing Car Carrier that is currently being developed under a grant from the Swedish Transport Agency.

The ship will need multiple wing sails, but in this study, a single sail without ship hull is considered. The single element wing sail has a planform area of 1844 m2, over a span of H = 80 m, with a mean chord length of 23 m. This gives an aspect ratio of 3.47. The wing has a taper ratio of 0.5, with a straight quarter-chord line and a curved leading and trailing edge. The section shape is a NACA0015 airfoil, which has been modified with a thin trailing edge.

Two variations of this condition are simulated, one with uniform inflow, and one with an atmospheric boundary layer including twist (change of apparent wind angle (AWA) and speed (AWS) over height).

For the uniform inflow case, a wind speed of 10 m/s is applied uniformly over the sail span. The wing sail is simulated at four different angles of attack (5, 10, 15 and 20 degrees).

For the twisted, atmospheric boundary layer case, the wind reference position is at 10 m above MSL. The case corresponds to a vessel sailing at a true wind angle (TWA) of 45 degrees, in a true wind speed (TWS) of 10 m/s and with a ship speed of 10 knots. The reference wind is used for determining the force direction and for non-dimensionalisation. The resulting apparent wind speed is 14.1 m/s, with an apparent wind angle of 30 degrees. To achieve the desired angle of attack, the sheeting angle is changed.

In order to account for ship freeboard height, the wing sail root is placed at 27 m above the water surface. This means that the wind velocity and angle vary considerably between the reference position and the wing root, and further between the wing root and the tip.

Thus, the values of the forces and force coefficients cannot be compared directly between uniform and twisted flow conditions. However, this treatment was considered most consistent with industry practice today, as used in the analysis of speed/power trial data [3], where the 10 m wind is always used as a reference.

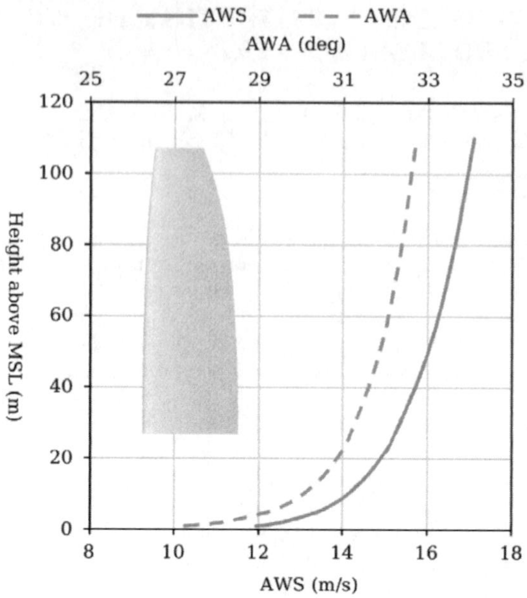

Figure 1: AWA and AWS as a function of height for the twisted flow case. The wing sail shows position and extent of the sail in relation to the wind profile.

2.2 ROTOR SAIL

A rotor sail (Flettner rotor) is a rotating cylinder placed vertically on the ship's deck. This WPS technology, which was originally invented in the 1930s, has recently seen a renaissance and has been retrofitted to a number of commercial vessels.

Even though interaction effects are important for performance, in this study a single rotor without ship hull is used to evaluate the suitability of the proposed methods for predicting single-rotor performance.

The rotor has a diameter D=3m and a height H=18m, giving an aspect ratio of AR=6. The tip of the rotor has an end disk of a diameter De=6m.

The ship (in this case the bottom of the CFD domain) is assumed to sail at 14 knots. Similar to the study for the wing sails, atmospheric boundary layer effect is considered and the resultant inflow to the rotor becomes a twisted velocity profile. The wind conditions and the corresponding AWS and AWA at the 10 m reference height, as well as the rotor RPMs, are given in Table 1.

Table 1: Inflow conditions for the rotor sail

$V_s = 14$ (kn)					
Case	**TWS (m/s)**	**TWA (°)**	**AWS (m/s)**	**AWA (°)**	**RPM**
1	8	90	10.7	48	200
2	10	200	5	222	170

3. METHODS

3.1 CFD SIMULATIONS

A full 3D CFD simulation is the most comprehensive approach. Assuming an adequate mesh is used, and suitable models are selected, the forces predicted by this approach is expected to have the highest accuracy of those presented in the paper. The cost of such a computation is however relatively high and still time-consuming, making it less suitable in a route simulation where the wind condition and the travelling route are dynamically varying. A limited number of CFD simulations provide benchmark data to assist the development of the more efficient methods discussed below.

In a full CFD simulation, the aerodynamic loads on the superstructure and the wing sail/rotor are calculated by solving the Reynolds Averaged Navier-Stokes (RANS) equations on a computational mesh in a pre-defined domain. Additional transport equations are used to model the turbulence of the bulk flow (turbulence closure). In the present study, Menter's SST k-ω model is adopted.

Simulation examples are given for the wing sail and the rotor below.

3.1 (a) 2D CFD Simulations of a NACA0015 Airfoil

Two of the simplified methods proposed depend on the availability of 2D airfoil data, in the form of lift, drag and pitching moment curves.

This data can be experimental if such data can be obtained for the airfoil in question, and over a relevant range of operating conditions. Due to the large scale of these vessels, the Reynolds number in full-scale operation easily reaches 10^7. For the wing and conditions considered here, the mean chord Reynolds number reaches 2.7×10^7. At such high Reynolds numbers, experimental data is difficult to obtain.

Considering the difficulty in obtaining experimental data, an alternative is to derive the 2D airfoil data using CFD. Since the simulations can be performed in 2D, the computational effort is low. This allows high-fidelity simulations, with highly dense grids and complicated models to be used even for parametric studies of wing sail section geometry. For example, the boundary layers can be resolved down to a low Y+ (<0.5), which allows effects such as separation and transition to be modelled.

Such high-fidelity modelling should be used in the design of a wing sail section. However, for this study, a lower-fidelity model, with a coarser grid and without transition modelling, has been used, in order to maintain similarity with 3D CFD. This is important to ensure a fair comparison of the simplified methods with 3D CFD. However, this negates one advantage of the SILL and NL-LL methods; they allow high fidelity viscous data to be used for VPP/route simulations.

For the 2D simulations performed here, a rectangular domain with a length of 795 m ($30c_{root}$) and a width of 530 m ($20c_{root}$) is used. The domain is meshed with a hexahedral mesh.

The mesh is on the leading edge is refined to a cell size corresponding to 800 cells per c. The cells gradually grow towards the trailing edge, reaching a maximum size corresponding to 100 cells per c. The prism layers give an average Y+ of 36. The total 2D cell count is 60×10^3. The 2D mesh in the near-field around the airfoil is shown in Figure 2

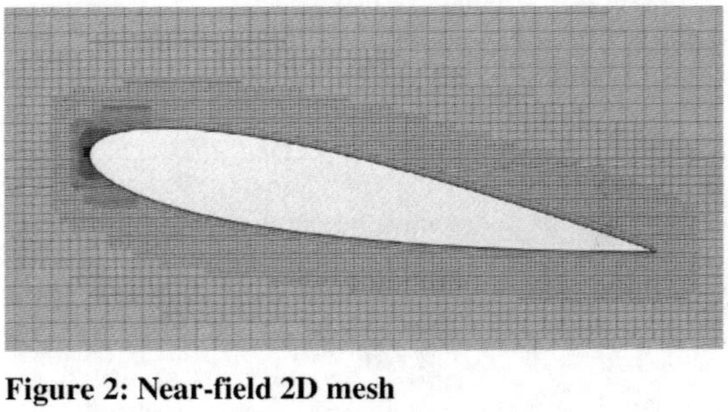

Figure 2: Near-field 2D mesh

3.1 (b) 3D CFD Simulations for the Wing Sail

3D simulations were performed to provide a reference case for the evaluation of the different methods. Such simulations were performed for both uniform and twisted flow. The simulations were set up to ensure maximum consistency between 2D and 3D CFD simulations

The 3D domain has the same X-Y dimensions as the 2D domain. The vertical (Z) dimension was 265 m.
The mesh settings are the same as those used in 2D, resulting in a total 3D cell count of 6.5×10^6.

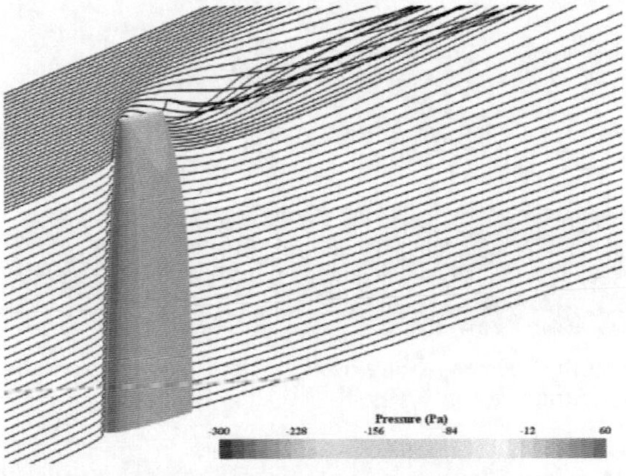

Figure 3: 3D CFD simulation of wing sail at α = 15° showing streamlines and tip vortex formation

3.1 (c) CFD Simulation for the Rotor Sails

The simulation was performed for a full-scale generic Flettner rotor with a Reynolds number Re $=2.5 \times 10^6$ based on rotor diameter and ship speed of 14 knots. A rectangular computational domain is defined, with the inlet boundary located at 5H and the outlet boundary at 8H distance to the rotor axis respectively. The rotor is standing on the bottom (representing the deck of the ship) where no-slip wall boundary condition is applied. The mesh is of hexahedral mesh type and consists of 3.3 million cells, with a $y^+=50$ for the non-dimensional wall distance of the first cell layer. The surface mesh of the rotor is shown in Figure 4.

Flow around a rotating cylinder is very complex due to the moving boundary layer caused by rotation. Complicated vortices and shedding phenomena are present. These features, in turn, affect the aerodynamic forces of the rotor in different ways. Figure 5 shows a snapshot of the vortices formed behind a rotor at a spin ratio SR=5. A thorough investigation of a single rotor performance can be found in our previous work [1].

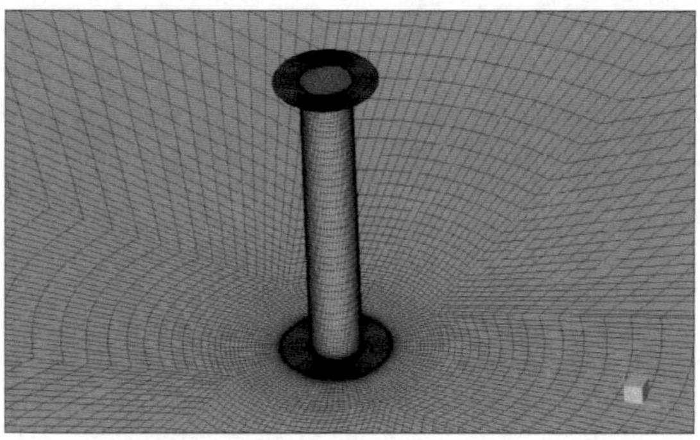

Figure 4: Surface mesh of the rotor

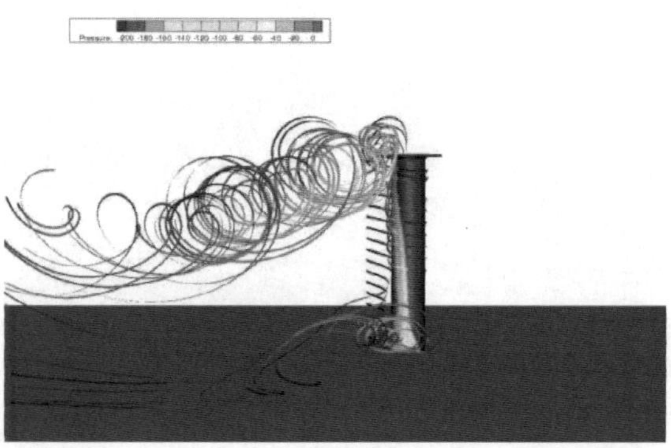

Figure 5: Example of various vortices downstream of the rotor

3.2 QUASI-3D METHODS

The quasi-3D methodology is used for extracting aerodynamic forces from the WPS device at a large number of flow conditions. The result is fed into the ship route simulation or a VPP.

The quasi-3D methodology presented here comes in several versions. Methods A and B are used for wing sails and method C for rotor sails. All methods make use of a limited number of CFD simulations, either 2D or 3D, from which the aerodynamic forces acting on the WPS device are extracted. These are then integrated to get the total forces and moments.

In the methods here called SILL and NL-LL, the sectional forces are based on 2D CFD (or experimental data), and the 3D effects are added using Lifting Line theory. In the 3D-SIM method, the sectional forces are extracted from 3D CFD simulations, and thus the 3D effects are already represented in the simulation. Due to the complex flow phenomena that occur on rotor sails, the SILL, NL-LL and VLM methods are not considered suitable for analysis of rotor sails.

3.2 (a) Sectional Integration with 3D-correction (SILL)

This method uses 2D airfoil lift and drag curves as input. In order to account for planform shape and wind twist, an integration procedure is performed in the vertical direction, considering the local wing area, wind speed and angle of attack, in order to determine global forces on the wing. The integrated forces are then corrected for 3D effects, using Prandtl's Lifting Line theory [7], assuming an elliptic distribution of lift.

Below the method will be described step-by-step:

1. Generate table describing C_L, C_D, and C_M for the 2D airfoil, as a function of angle of attack α
2. Discretise the 3D wing into a finite number of sections in the vertical (Z) direction
3. For each sail section:
   a. Compute apparent wind from TW, V_S and ABL
   b. Compute angle of attack considering sheeting angle and local (twisted) apparent wind
   c. Compute local lift and drag forces and pitching moment from interpolated 2D table values
4. Vertically integrate local forces and moments
5. Compute mean apparent wind over the wing span
6. Use Lifting-Line theory to compute induced drag and loss-of-lift in an "equivalent" aerodynamic coordinate system defined by the mean AOA over the wing, giving global 3D forces for the wing

This method represents a computationally efficient way to estimate forces and moments on a 3D wing, based on 2D airfoil data. The most significant simplification is the assumption of an elliptic lift distribution, which allows the 3D effects to be computed with analytical formulas.

3.2 (b) Non-linear Lifting Line algorithm (NL-LL)

The method is best described as a non-linear Lifting Line algorithm. The main difference to the SILL method is that the lift distribution is not assumed to be elliptic, but instead solved iteratively. Many different implementations have been proposed, see [5].

The method used here is that proposed by [7]. It is commonly called a Γ-coupled iterative algorithm. This version offers a simple implementation, but is numerically unstable, meaning that large under-relaxation factors must be employed, and thus, a considerable number of iterations have to be completed to reach convergence.

The instability of the iterative procedure algorithms can also result in poor convergence (or even divergence) at stall and post-stall angles of attack.
Below the method will be described step-by-step:

1. Discretise wing into a finite number of stations
2. Assume a span-wise distribution of circulation Γ
3. Compute the induced angle of attack (α_i) for each station from the circulation distribution
4. From a known geometric angle of attack α, determine the effective angle of attack $\alpha_e = \alpha - \alpha_i$
5. Obtain the local viscous lift coefficient $C_{l,v}$ from the 2D tabular data, at each station, using α_e
6. Compute a new circulation distribution using the Kutta-Joukowski theorem $\Gamma = 1/2 \cdot V \cdot c \cdot C_{l,v}$
7. Update circulation distribution $\Gamma = \Gamma_{old} + d(\Gamma_{new} - \Gamma_{old})$ where d is the under-relaxation factor (≈0.05)
8. Iterate over steps 3-7 until $|\Gamma_{new} - \Gamma_{old}| \leq \varepsilon$ where ε is a pre-defined convergence criterion ($\approx 1e^{-4}$)

3.2 (c) 3D Sectional Integration Method (3D-SIM)

The 3D-SIM method is used for the rotor sail. The procedure to calculate propulsive forces consists of the following steps:

1. Perform 3D CFD simulations of the rotor at a range of spin ratios (see 3.1 (c)) in uniform flow. Divide the rotor in a number of sections in the span-wise direction and extract the local sectional lift and drag in the coordinate system of the local inflow. Save them in the form of sectional coefficient, C_l and C_d.
2. For a given ship speed and an arbitrary true wind in a route simulation, determine first the Apparent Wind Speed (AWS) and Apparent Wind Angle (AWA) at each section
3. Compute the local spin ratio, lift and drag per section using sectional coefficient C_l and C_d.
4. Project the force in the x and y direction to obtain the local thrust (or drag) and side force.
5. Obtain the rotor propulsive force F_{Rx} and side force F_{Ry} by integrating the span-wise sectional forces.

3.3 VORTEX LATTICE METHOD (VLM)

For the wing sail case, estimations of the lift and drag coefficients were also done using a 3D Vortex Lattice Method (VLM) code. A 3D panel method as used assumes potential flow only and hence does not require previously computed CFD simulations.

The code is based on the theory developed in [10] with vortex rings as singularity elements. A 3D-VLM code enables to compute aerodynamic coefficient for wings or sails, accounting for three-dimensional effects. However, it assumes inviscid flow, thus neglecting any viscous effect. For small angles of attack, and if the flow remains attached, the results are expected to be reasonably correct, but detached flows cannot be represented. The drag forces are expected to be underestimated since VLM methods only account for induced drag.

In brief, the wing geometry is represented by a small number of rectangular panels that are placed on the wing's camber, i.e., in the case of a symmetrical airfoil section, the wing is represented as a flat plate. On each of these panels, a vortex ring is placed, with the leading-edge segment located at the quarter-chord of the panel. The circulation on the lattice is calculated by imposing a boundary condition of zero flow through the surface at the collocation point (located at three-quarter chord and mid-height) of the panel.

After the wing's trailing edge, an additional row of panels and lattices is added to represent the wake. The lattices in the wake have the same vortex strength as the last panel of the wing in order to satisfy the Kutta condition of zero circulation at the trailing edge. The wake can be modelled with either a fixed wake model or using a wake relaxation routine. In the fixed wake model, the row of wake panels is aligned with the freestream velocity so that the wake does not generate lift, and the panels are extended far behind the wing.

The wake relaxation routine is based on the time-stepping method described in [8]. For several pseudo-time-steps, a set of span-wise vortex rings is released in the wake, with an intensity equal to that of the last vortex ring on the wing. The vortices are then aligned with the local flow velocity to make them force-free. The pseudo-time-steps is continued until the strength of the vortices converges.

The code used is based on [9] with modifications to allow for multiple wings calculations parametrised study. More details about the implementation and about the VLM can be found in [9].

For the present paper, the wing is discretised in eight panels in the chordwise direction and nine in the span-wise direction. The wake is modelled as a fixed wake expanding about five chord lengths behind the wing, as can be seen in Figure 6.

To choose the number of panels, different levels of discretization were evaluated and the minimum number of panels that ensures grid-independency was taken. The difference in lift coefficient between the chosen grid of 72 panels and a grid with 560 panels is about 1%. Similarly, the choice of a fixed wake is done after comparing the results.

For the same grid, the difference between the lift coefficient with a fixed or free wake is about 1%. An example of the pressure distribution on the wing for the case of uniform flow at an angle of attack of 10° is shown in Figure 7.

For comparison with the other methods, the calculations for one angle of attack takes less than 1s on a standard laptop.

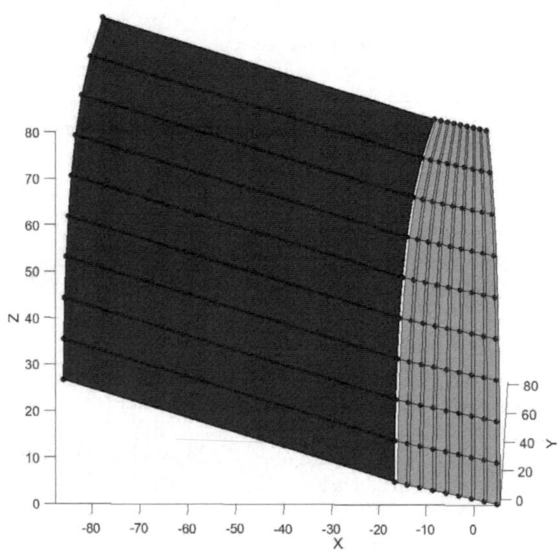

Figure 6: Discretization of the wing in 8x9 panels. The geometry is in grey, while the blue dots show the corners of the vortex lattices and the red panels represent the wake. The axes are in meters.

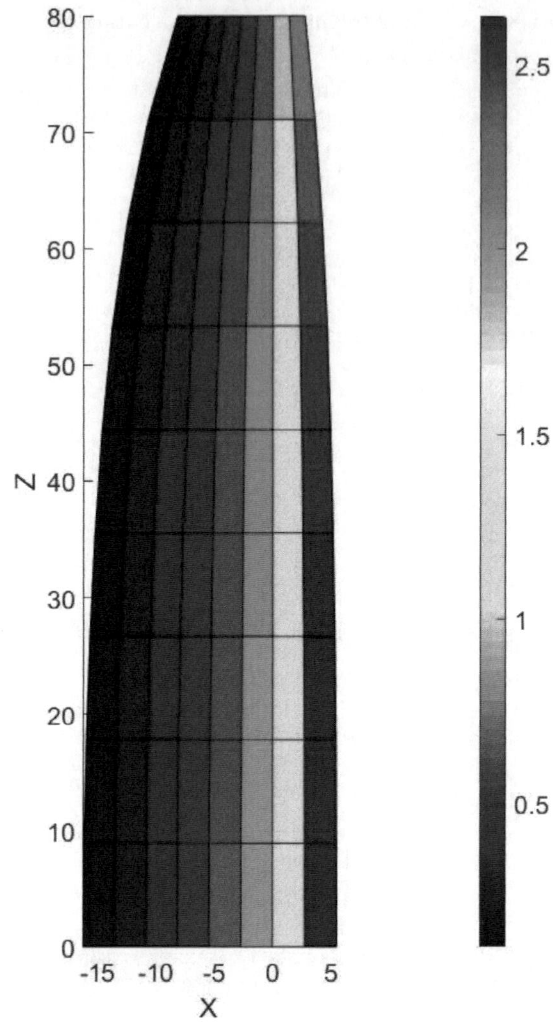

Figure 7: Pressure distribution on the wing for a uniform flow at 10° AOA. The axes are in meter.

4. RESULTS

In this section, the ability of the alternative methods to predict the aerodynamic forces is discussed. For the wing sail, results from 4 methods are presented, and for the rotor sail, there are results from 2 methods.

4.1 WING SAIL

For the wing sail, the 3D lift and drag coefficients (C_D and C_L) predicted by the different methods are compared in Figures 8 to 11. In addition, the span-wise distribution of the local lift coefficient C_l is presented in Figure 12 and 13 for three of the methods (3D CFD, NL-LL and VLM). For the strip integration method (SILL), no such information is available, since the effect of the 3D lift distribution is implicitly considered through Prandtl's formulas, and not explicitly computed.

A comparison of these local quantities allows the predictive ability of these methods to be evaluated, and it may help to explain the differences found between the methods regarding the global forces. In these comparisons, the results from 3D CFD is considered to be the reference, since all viscous and 3D effects are represented. However, it does not mean that any of the results can be considered validated.

4.1 (a) Uniform Flow

The VLM underpredicts the drag C_D at all angles of attack. This is expected since the VLM completely neglects viscous effects, and these are at least partly accounted for in the other methods.

In the pre-stall region, SILL consistently over-predicts C_D, while the NL-LL agree well with 3D CFD. At α = 15° (approaching stall threshold) the trend is reversed, and NL-LL starts to over-predict C_D, while SILL shows good agreement with 3D CFD.

At α = 20°, when the wing has started to stall at the root, all three simplified methods under-predict the drag coefficient. In comparison with 3D CFD, NL-LL gives the best prediction, with a difference of -9.8%. The VLM predicts the lowest drag, with a difference of -27%.

For the lift coefficient, C_L, the three simplified methods show good agreement with 3D CFD, up to and including α = 15°. However, in the post-stall region (α = 20°), VLM and NL-LL over-predict C_L, i.e. predicting no stall.

This is expected for the VLM, which is based on thin airfoil theory, and thus assumes a linear relationship between lift coefficient and angle of attack. The NL-LL should be able to predict the loss-of-lift at post-stall angles since this is included in the 2D airfoil L/D curves. However, as has been shown by [5] and [6], the failure to predict post-stall behaviour is a result of the numerical instability of the coupling algorithm implemented in the NL-LL.

3D CFD and SILL agree well, even for post-stall angles. This indicates that post-stall behaviour is captured in the 2D simulations and that the assumption of an elliptical lift distribution has a negligible effect for this case.

Figure 12 shows the distribution of the local lift coefficient C_l in the span-wise direction of the wing, as predicted by three of the methods, for uniform flow. This can help explain why there is a difference between the forces predicted by the methods.

For angles of attack α = 5-15°, Figure 12 shows that, in comparison to 3D CFD, NL-LL method over-predicts the local lift coefficient over a large extent of the span. For α = 5-10°, the VLM under-predicts the local C_l at the root but over-predicts the local C_l towards the tip. This indicates a more tip loaded wing. At α = 15°, the VLM agrees very well with 3D CFD, only showing a slight over-prediction towards the tip.

For post-stall (α = 20°) there is as expected, given the difference in predicted force coefficients, a large difference between the predicted distributions. The VLM and NL-LL over-predicts the local lift coefficient for all of the span, but in particular at the wing root. In the 3D CFD results, the low local C_l at the root indicates that the wing root is stalling. However, the VLM and the NL-LL fails to capture this entirely.

4.1 (b) Twisted Flow

Overall, the methods are considerably more scattered for the twisted flow case. NL-LL over predicts C_D across all angles of attack. The VLM over predicts up to α = 15°, but slightly under predicts post-stall (α = 20°), by -10%. The NL-LL method as well as SILL over-predict C_D considerably post-stall, by 16% and 29% respectively.

For C_L, the 4 methods agree well up to α = 10 deg. At α = 15°, the three simplified methods over-predict lift. SILL over-predicts slightly (6.1%), while NL-LL and VLM give similar results, over-predicting with ≈16%. At α = 20°, NL-LL and VLM again over-predict (≈59%), while SILL now under-predicts considerably (37%).

Figure 13 shows the distribution of the local lift coefficient C_l in the span-wise direction of the wing, as predicted by three of the methods, for twisted flow.

For angles of attack α = 5-10°, both VLM and NL-LL under-predicts the C_l at the root, but over-predicts at the tip (again, indicating higher tip loading). This will most likely lead to a higher induced drag being predicted and thus may explain the difference in the predicted drag coefficient.

For α = 15-20°, the same behaviour as for uniform flow can be seen, where the VLM and NL-LL dramatically over-predict the local lift coefficient C_l at the root, indicating that the methods fail to predict stall.

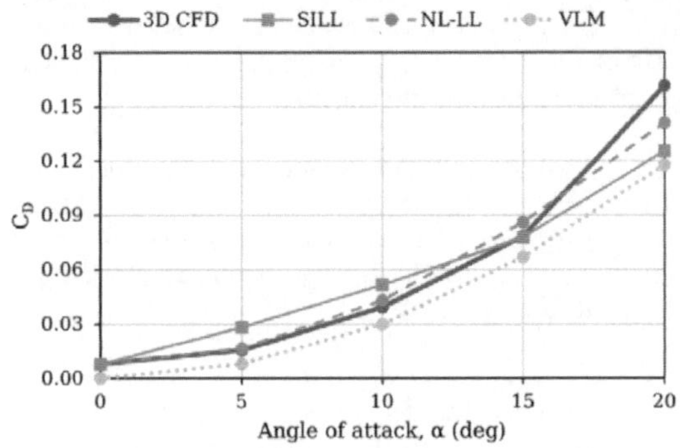

Figure 8: Predicted drag coefficient C_D, uniform flow

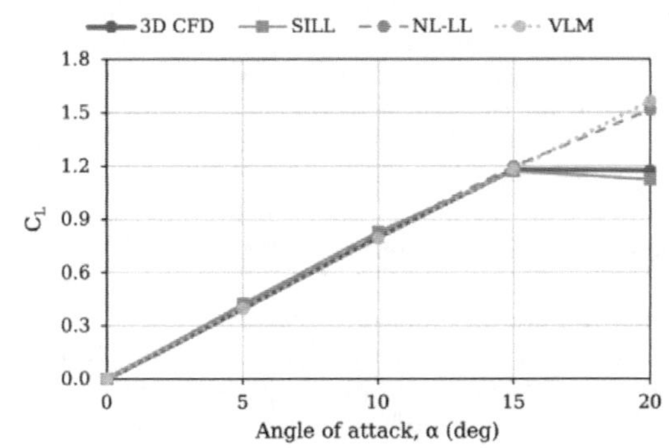

Figure 9: Predicted lift coefficient C_L, uniform flow

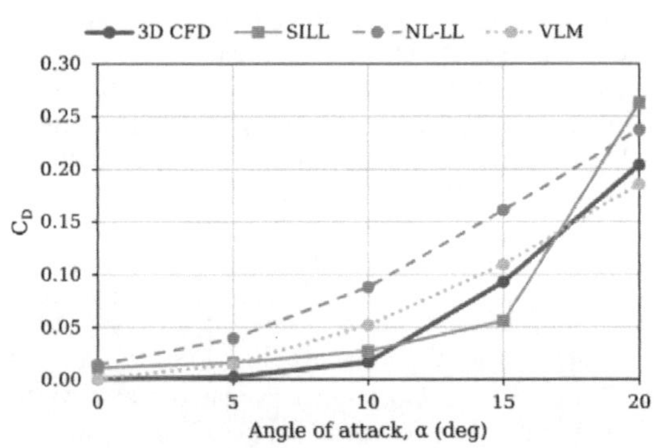

Figure 10: Predicted drag coefficient C_D, twisted flow

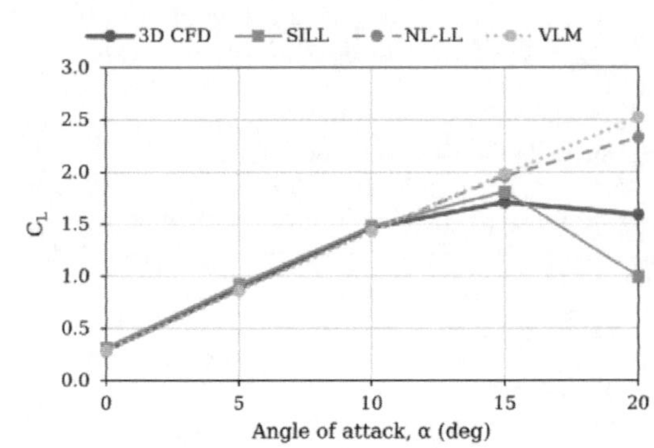

Figure 11: Predicted lift coefficient C_L, twisted flow

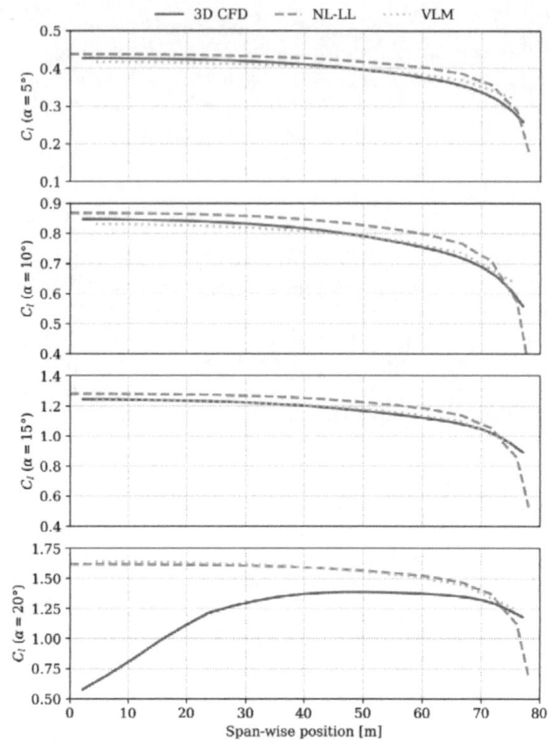

Figure 12: Local lift coefficient over the wing for uniform flow and four angles of attack

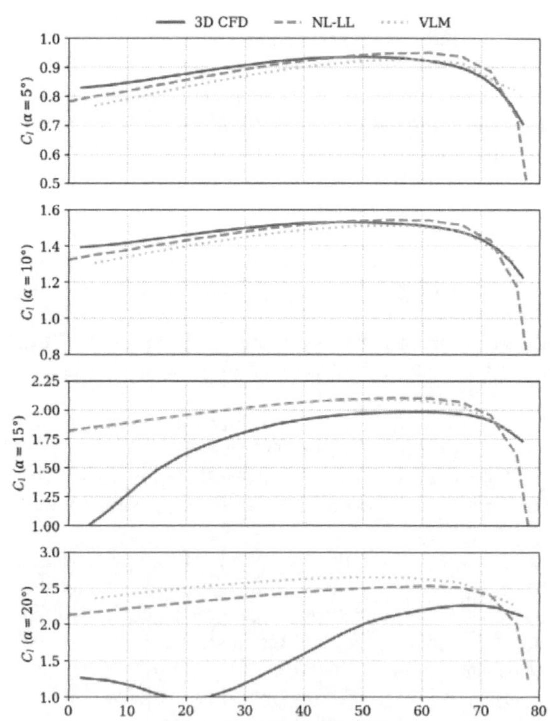

Figure 13: Local lift coefficient over the wing for twisted flow and four angles of attack

4.1 (c) Effect on Performance Prediction

The differences in predicted sail forces produced by the methods described above result in different predictions of the overall ship's performance. In order to illustrate the sensitivity of the performance prediction (and thus, the subsequent route simulation) to errors in the aerodynamic force predictions, VPP computations have been performed for a generic wind assisted cargo vessel concept with four identical wing sails. The computations have been performed for TWS = 10 m/s.

Figure 14 shows the ship speed predicted by the VPP, using aerodynamic data corrected for the error obtained with the different methods, in uniform inflow conditions. The Sectional Integration Lifting Line method (SILL) predicts ship speeds close to the full 3D CFD predictions across all true wind angles. The VLM and the Non-Linear Lifting Line (NL-LL) method overpredicts the ship speed, especially at beam reach, where the sails are likely to be sheeted close to stall, where the VLM and NL-LL gives a considerable error.

Figure 15 shows the ship speed predicted in twisted flow conditions, where again the SILL method results in a predicted speed that is similar compared to 3D CFD, across all true wind angles. The VLM and NL-LL now result in under-prediction of the ship speed over a large part of the true wind angle range, except for deep broad reach and running wind angles.

Tables 2 and 3 show the predicted speed in knots, for true wind angles of 45 and 90, respectively. Predictions are shown for each method, and for uniform flow conditions as well as twisted flow condition. Please note that the predicted speeds cannot be directly compared between uniform and twisted flow since the definition of the reference wind along with the atmospheric boundary layer means that the average wind speed over the sail span is considerably higher for twisted flow conditions than for uniform flow conditions.

Table 2: Predicted ship speed at TWA=45

Method	V_S (kn)	
	Uniform flow	Twisted flow
3D CFD	10.5	14.2
SILL	10.5	14.3
NL-LL	11.4	9.1
VLM	21.1	11.2

Table 3: Predicted ship speed at TWA=90

Method	V_S (kn)	
	Uniform flow	Twisted flow
3D CFD	15.9	20.8
SILL	15.8	20.3
NL-LL	17.7	18.2
VLM	18.1	19.1

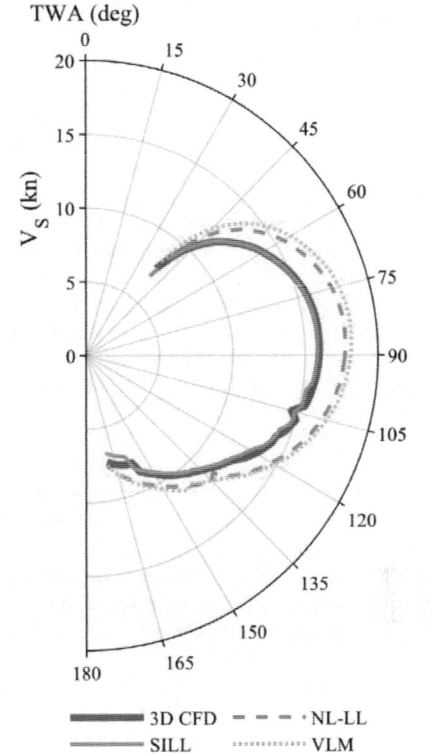

Figure 14: Velocity prediction for uniform flow

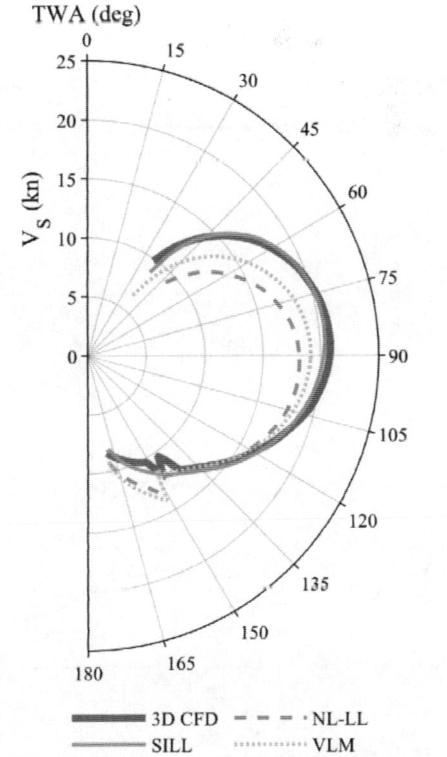

Figure 15: Velocity prediction for twisted flow

4.2 ROTOR SAIL

To demonstrate the reliability of 3D-SIM, both the full CFD simulation and the 3D-SIM method are used to predict aerodynamic forces on the rotor sail for two wind conditions as described in Section 2.2. The span-wise distribution of sectional forces predicted by the two methods is compared in Figure 13 for the two cases. The sectional forces are expressed as local coefficient C_x and C_y and plotted against the non-dimensional rotor height h/H. As seen in the figure, the agreement between the two methods is very good for Case 1 and quite good for Case 2, especially in the midsections. At the tip and root of the rotor, the agreement is less satisfactory. This is related to the complexity of the tip vortices and the necklace vortices at the root.

The data from the CFD simulation is extracted from a particular time step whereas the 3D-SIM prediction is obtained from averaged sectional coefficients. It is known that at certain spin ratios the vortices at the tip and root are highly dynamic, which makes the C_l and C_d more unstable for the end sections than the midsections, leading to a notable deviation between the two methods.

Compared with the total thrust (F_{Rx} force) predicted by the CFD simulation, the 3D-SIM method gives a difference of 1.1% in Case 1 and 4.2% in Case 2. The difference between the side force (F_{Ry}) is 1.3% in Case 1 and 3.6% in Case 2. This data is also given in Table 3.

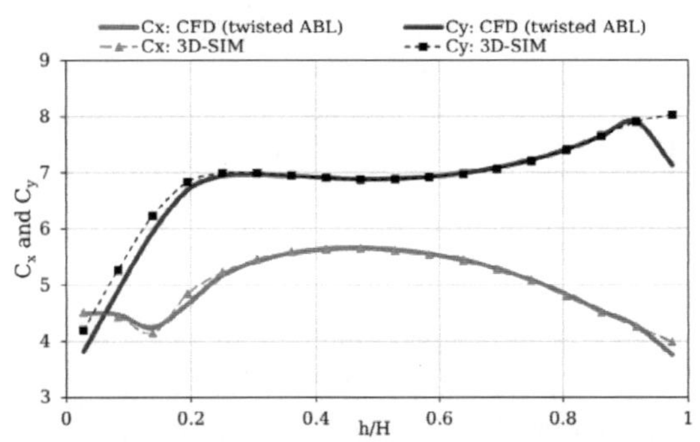

(a) Case 1

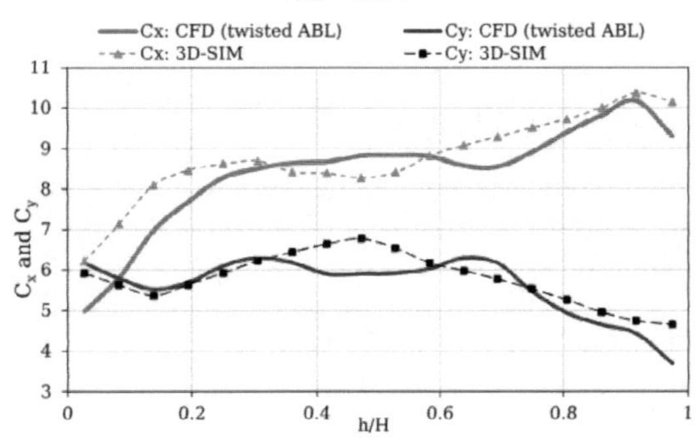

(b) Case 2

Figure 16: Local rotor force coefficients predicted by CFD and 3D-SIM, for Case 1 (a) and 2 (b)

4.2 (a) Effect on Fuel Saving Prediction

To illustrate the impact of the error due to the simplification in the Sectional Integration method on the predicted fuel consumption on a route, we assume a generic ship where the rotor contributes to 20% of the total thrust in Case 1 and 25% in Case 2. We also assume that the ship propeller (at same ship speed) has the same propulsive efficiency in the two cases. We calculate the total power generated by the propeller and the rotor, and estimate the fuel consumption of the ship in a 24-hour's journey with the wind conditions given in Case 1 and 2. Since we are only interested in the deviations introduced by the 3D-SIM method, we present only the difference in the fuel consumption predicted by the 3D-SIM method relative to that by the CFD method in Table 3.

To explain the data in Table 2, the 3D-SIM method will lead to a 0.2% deviation in the prediction of fuel consumption in Case 1, compared with the CFD simulation. The method will result in a 1.1% difference in fuel consumption in Case 2 compared with the 3D CFD simulation. The assumption of a wind speed TWS=8~10 m/s and a 20~25% of total thrust coming from the rotor during the entire route makes the present assessment quite optimistic for the thrust contribution of the rotor.

In a real-time route simulation, the thrust contribution of rotor may not be as such high all the time and the deviations introduced by the 3D-SIM method will be even less. Therefore, it can be concluded that the deviation introduced by this method will not lead to a significant error in the estimation of fuel consumption.

Table 4: Difference introduced by the 3D-SIM method compared with the CFD simulation

	Rotor/total thrust ratio	Diff. in thrust	Diff. in side force	Diff. fuel consumption
Case	**T_R/T (%)**	**DF_{Rx} (%)**	**DF_{Ry} (%)**	**DF (%)**
1	20	1.1	1.3	0.2
2	25	4.2	3.6	1.1

5. CONCLUSIONS

Performance predictions of WPS solutions require the aerodynamic forces to be predicted at a large number of flow conditions. With 3D CFD simulations, it is possible to capture the dominant flow physics, but these computations are very time-consuming. A feasible approach that combines 3D CFD, 2D CFD and lifting line theory has been presented here. The idea is to divide the device into segments and derive the aerodynamic forces for each segment from either 2D or 3D CFD simulations. The total forces can then be integrated; hence, the methodology is named Sectional Integration method. When 2D data is used for the sectional forces, the induced drag and loss of lift is derived from with linear lifting line (assuming elliptic lift distribution) or non-linear lifting line.

The methods are demonstrated for two types of WPS: a wing sail and a rotor. For the wing sail, an inviscid Vortex Lattice Method is also used for comparison.

5.1 WING SAILS

For angles of attack up to 10 degrees, all methods predict the lift well. The drag is harder to capture accurately, especially when a twisted inflow from an atmospheric boundary layer is applied.

Of the tested simplified methods, the Sectional Integration with Lifting Line correction (SILL) predicts the lift and drag best, with a small difference up to close stall (15 degrees angle of attack for this case). The resulting upwind ship speed is predicted within 0.5 knots for twisted flow conditions, compared to when using aerodynamic forces from 3D CFD. Using VLM or Non-Linear Lifting Line the ship speed is not well predicted, with a maximum error of 5.1 knots.

For angles of attack close to stall (20 degrees in this case), none of the simplified methods are found to be reliable.

It is concluded that the SILL method can be used in combination with 3D CFD simulations in the concept design phase of a ship with wing sails, as long as the design range is limited to pre-stall angles of attacks.

5.2 ROTOR SAILS

The differences in the rotor force predicted by the 3D-SIM method are less than 4.3% as compared with a full 3D CFD simulation for the two studied cases. The error introduced in the estimation of fuel consumption due to these differences is 3.6% or less. The study demonstrates that the 3D-SIM method is a viable tool for route simulations and it will not lead to a significant error in the estimation of fuel consumption.

6. FUTURE WORK

In the computations using twisted inflow, the atmospheric boundary layer is modelled with a power law (exponent 1/7) over the entire span of the WPS (up to 100m above MSL). The valid range of this model has not been investigated. Measurements of the velocity gradient over open sea at this range of altitudes should be performed to verify this.

The 3D-SIM method has been tested in a limited number of cases, further study in more variable wind conditions may be necessary to have a complete assessment of the method.

The Non-Linear Lifting Line method could possibly be improved using a closely coupled algorithm as suggested in [5]. This will be investigated in future work.

Whether it is necessary to consider the atmospheric boundary layer and the subsequently induced wind twist when evaluating the performance of WPS is another subject that will be investigated in detail in future studies.

7. ACKNOWLEDGEMENTS

This work was supported by the Swedish Transport Agency under grant TRV 2018/96451 and Fru Martina Lundgrens fond för sjöfartsteknisk forskning.

A subset of the computations were performed on resources at Chalmers Centre for Computational Science and Engineering (C3SE) provided by the Swedish National Infrastructure for Computing (SNIC).

8. REFERENCES

[1] Li, D-Q., Leer-Andersen, M., and Allenström, B., 'Performance and vortex formation of Flettner rotors at high Reynolds numbers', *Proc. of 29th Sym. On Naval Hydrodynamics*, 2012.

[2] Froude, W., 'On the Rolling of Ships', *Transactions of RINA*, 1861.

[3] ITTC, 'Preparation, Conduct and Analysis of Speed/Power Trials', ITTC Recommended Procedures and Guidelines 7.5-04-01-01.2, 2017.

[4] IMO, '2014 guidelines on survey and certification of the energy efficiency design index (EEDI), as amended (resolution mepc.254(67), as amended by resolution mepc.261(68))', 2015.

[5] Gallay, S., and Laurendeau, E., 'Nonlinear Generalized Lifting-Line Coupling Algorithms for Pre/Poststall Flows', AIAA Journal, Vol. 53, No. 7, 2015, pp. 1-19

[6] Graf, K., Hoeve, A.V., Watin, S., 'Comparison of full 3D-RANS simulations with 2D-RANS/lifting line method calculations for the flow analysis of rigid wings for high performance multihulls', Ocean Engineering, No. 90, 2014, pp. 49-61

[7] Anderson, J. D., 'Fundamentals of aerodynamics', 5th ed., McGraw-Hill, 2005

[8] Katz, J., and Plotkin, A. 'Low-speed aerodynamics: from wing theory to panel methods, McGraw-Hill, 1991

[9] Helmstad, A., Larsson, T., 'An Aeroelastic Implementation for Yacht Sails and Rigs.' Master's thesis, KTH Royal Institute of Technology, 2013

9. AUTHORS BIOGRAPHY

Adam Persson is a PhD student at SSPA Sweden AB. His research aims to develop methods for prediction of the dynamics of sailing vessels. He has previously worked with CFD simulations as well as experimental methods on this topic. He has also been working with naval architecture for an America's Cup team.

Da-Qing Li works in the Research Department at SSPA Sweden AB. His experience includes CFD methods and their applications in propeller/propulsion, multiphase flow (free surface and cavitation) and underwater radiated noise.

Fredrik Olsson works in the Maritime Consulting Department at SSPA Sweden AB. He is the core developer of the simulation software suite SEAMAN. His previous experience includes simulation and modelling of non-linear dynamical systems, quasi-static thrust optimisation/allocation problems, route optimisation as well as general manoeuvring and seakeeping simulation.

Sofia Werner holds the current position of Strategic Research Hydrodynamics at SSPA. She is responsible for development within fluid dynamics at SSPA. Her previous experience includes towing tank testing and ship design using CFD. She is currently Chairman of the ITTC Specialist Committee for CFD and EFD Combined Methods.

Ulysse Dhomé is a PhD student at the Centre for Naval Architecture at KTH Royal Institute of Technology, Stockholm. His previous experience includes experimental and numerical work on sail aerodynamics.

THE EFFECT OF LEEWAY ANGLE ON THE PROPELLER PERFORMANCE

J. J. A. Schot and R. Eggers, Maritime Research Institute (MARIN), the Netherlands

SUMMARY

One of the aspects influencing the performance of wind assisted vessels is the effect of leeway (drift) angle on the propeller performance. It is known that due to leeway the delivered propeller power and propulsive efficiency can vary significantly from the straight sailing condition. This effect of leeway angle is studied using a combination of viscous flow calculations and captive model tests for one twin screw and three single screw vessels. It is observed that the changes in mean axial velocity and pre-swirl rotation in the wakefield due to a combination of leeway angle and propeller suction are sufficient to describe the trends observed in captive model tests. This knowledge is used in a proposed prediction method to model the changes in propeller thrust and torque due to leeway angle at the design stage. The prediction model combined with a fit of the average wake parameters for the studied vessel types is finally used the show the trends in propulsive efficiency and delivered propeller power at constant propeller rotation rate and ship speed for small leeway angles.

NOMENCLATURE

V_S	Ship speed (kn, m/s)
β	Leeway / drift angle (deg)
u	Velocity along ship x-axis (m/s)
v	Velocity along ship y-axis (m/s)
J_V	Apparent advance coefficient (-)
J	Advance coefficient (-)
V_A	Advance speed of the propeller (m/s)
V_{AX}	Mean axial velocity at propeller (m/s)
K_T	Thrust coefficient (-)
K_Q	Torque coefficient (-)
T	Propeller thrust (kN)
Q	Propeller torque (kNm)
D	Propeller diameter (m)
n_p	Propeller rotation rate (Hz)
$n_{pre\text{-}swirl}$	Pre-swirl rotation rate (Hz)
w_{AX}	Axial wake fraction (-)
$n_{pre\text{-}swirl}'$	Dimensionless pre-swirl rotation (-)
w_T	Taylor wake fraction (-)
t	Thrust deduction factor (-)
η_D	Propulsive efficiency (-)
η_O	Open-water efficiency (-)
η_R	Relative rotative efficiency (-)
η_H	Hull efficiency (-)
P_D	Delivered propeller power (kW)

1. INTRODUCTION

One of the aspects influencing the performance of wind assisted vessels is the effect of leeway angle on the propeller performance. Besides forward thrust, wind propulsion devices also generate a side force which has to be balanced by the hull and rudder. As a result the ship will sail at a non-zero leeway angle and non-zero rudder angle. From observations in manoeuvring model tests and performance monitoring data it is known that due to the resulting leeway the propeller absorbed power can vary significantly from the straight ahead condition. Apart from this change in the propeller operating point also the propulsive efficiency can change.

Using a combination of data from viscous flow calculations and captive model tests, the main flow features affecting the propeller thrust and torque when sailing under a leeway angle are investigated. Based on previous research at MARIN in the context of Energy Saving Devices (ESD) [1] [2] it is expected that the mean axial velocity and mean rotation in the wakefield, often called pre-swirl, are the most important contributions affecting this aspect of the propeller performance.

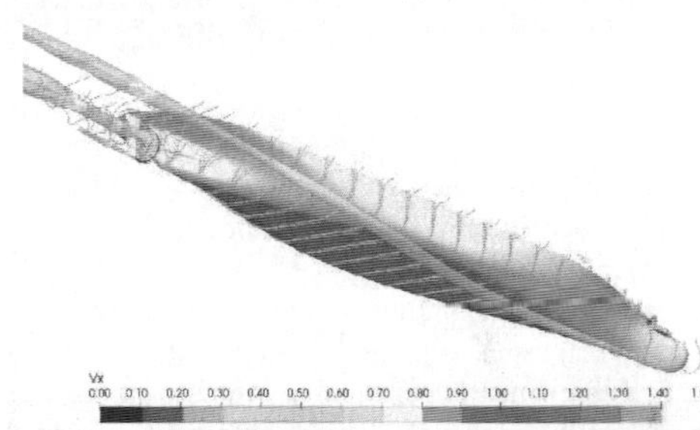

Figure 1: Example of the flow towards the propeller plane for the KCS container ship sailing at β=8°.

It is therefore studied if the changes in mean axial velocity and pre-swirl resulting from changes in the flow due to a combination of leeway angle and propeller suction (see Figure 1) are sufficient to explain the trends in thrust and torque measured during captive model tests. From this measurement the propeller-hull interaction coefficients (w_T, η_R) can be determined. These coefficients are known from research on ESDs to lose their meaning in describing physics of the propeller-hull interaction when pre-swirl is present. They are therefore only used for the prediction of the propeller performance using the open-water diagram and in order to compare to the captive model test results.

The measured results are compared to a proposed prediction method for the propeller thrust and torque of a vessel sailing at a leeway angle. This method uses a fit of the mean axial velocity in terms of the wake fraction (w_{AX}) and the mean pre-swirl rotation rate ($n_{pre\text{-}swirl}$) in the wakefield in combination with the open-water diagram. These values can be related to w_T, η_R which can be compared to the captive model tests in order to determine the accuracy of the used assumptions and prediction model.

The prediction model together with the fits of w_{AX} and $n_{pre\text{-}swirl}$ over leeway angles is finally applied to show the trends of the propulsive efficiency (η_D) and delivered propeller power (P_D) for the studied vessel types at constant propeller rotation rate and ship speed.

2. THEORY

2.1 EFFECT OF LEEWAY ON WAKE FIELD

For a vessel sailing under a leeway angle, the speed in the sailing direction (V_s) differs from the velocity along the ships x-axis (u). In addition a side velocity is present (v). The relation between u, v and V_s depending on the leeway angle (β) are given in equation 1 and 2. The used axis system definition is shown in Figure 2.

$$u = \cos(\beta)\ V_s \quad (1)$$

$$v = \sin(\beta)\ V_s \quad (2)$$

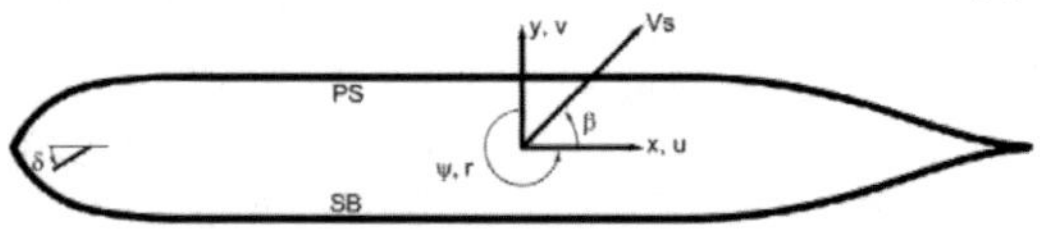

Figure 2: Axis system definition for leeway seen from above

The velocities in the propeller plane are generally reported relative to the ship speed (V_S) in the wakefield. The relative difference of the mean axial velocity (V_{AX}) in the propeller plane with the ship speed is called the wake fraction as defined in Equation 3. In order to differentiate with wake fractions obtained using other methods discussed later it is more specifically called the axial wake fraction (w_{AX}). The integral used to obtain w_{AX} from the wakefield is given in Equation 4.

$$w_{AX} = 1 - \frac{V_{AX}}{V_S} \quad (3)$$

$$w_{AX} = 1 - \frac{1}{V_S}\left[\frac{\int_{d/2}^{D/2}\int_0^{2\pi} V_x(r,\theta)\ r\ d\theta\ dr}{\int_{d/2}^{D/2}\int_0^{2\pi} r\ d\theta\ dr}\right] \quad (4)$$

Even in the case with no flow disturbance from the hull, w_{AX} will increase with increasing leeway angle due to the definition given in Equation 1. When the effect of the hull is considered a further decrease of the axial velocities and therefore increase of w_{AX} is present due to the shadowing effect of the hull. This effect is depending on the leeway angle.

The side velocity (v) due to drift is not considered directly when predicting the propeller performance, as its effects on the propeller thrust and torque are expected to average out over a propeller revolution [1]. However, it can contribute to other effects such as bearing loads, rudder forces and cavitation. This is not considered during this study, but can be determined using more complex calculation methods.

The side velocity however does however affect the studied parameters in the wakefield. The side force of the hull generated due to the side velocity is accompanied by the formation of vortices as shown in Figure 1. In the figure the relative axial velocity is shown on slices and in the propeller plane and the vortices forming around the hull are indicated in grey. If these vortices reach the propeller plane a significant amount of rotation is observed in the wakefield. An example of this effect on the wakefield of a container ship is shown in Figure 3 at model scale. The relative axial velocity is shown and the arrows indicate the relative radial and tangential velocity component.

The rotation in the wakefield is often called pre-swirl and it is known to affect the propeller thrust and torque. The average rotation rate due to pre-swirl is named $n_{pre\text{-}swirl}$ in this study and made non-dimensionless using the definition given in Equation 5. The non-dimensionless quantity can be used for scaling to other ship speeds and from model to full scale. The integral used to obtain $n_{pre\text{-}swirl}$ from the wakefield for a right turning propeller is given in Equation 6. For a left turning propeller a minus sign should be applied to the equation.

$$n_{pre-swirl}' = \frac{n_{pre-swirl}\ D}{V_s} \quad (5)$$

$$n_{pre-swirl} = \frac{\int_{d/2}^{D/2}\int_0^{2\pi} \frac{-V_t(r,\theta)}{2\pi r}\ r\ d\theta\ dr}{\int_{d/2}^{D/2}\int_0^{2\pi} r\ d\theta\ dr} \quad (6)$$

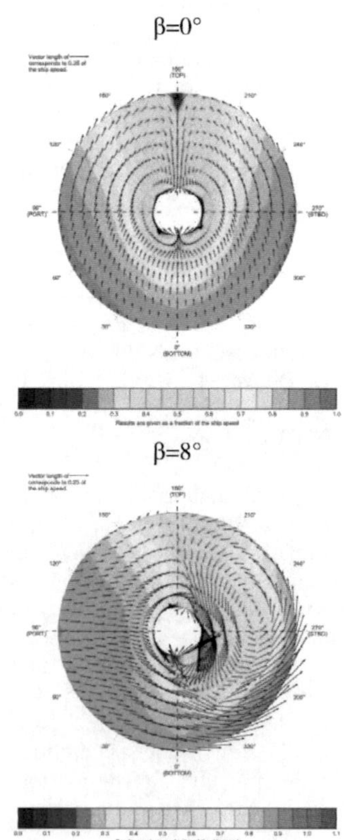

Figure 3: Example of the effect of drift on the wakefield for the KCS container ship at model scale.

2.2 PROPOSED PREDICTION MODEL

It is expected that the change in propeller trust and torque when sailing at a leeway angle is mainly caused by a mean axial velocity change (w_{AX}) and pre-swirl ($n_{pre\text{-}swirl}$) in the wakefield. In this paragraph the proposed prediction method for the propeller thrust and torque is explained which uses an estimate of these quantities over leeway angles in combination with the open-water diagram of the propeller.

The thrust and torque coefficient (K_T, K_Q) calculated using this method can be related to propulsive interaction coefficients (w_T, η_R) which differ from the straight-sailing condition. These predicted interaction coefficients are compared to captive model tests in chapter 3 to show that indeed the change in mean axial velocity and pre-swirl is sufficient to explain the observed trends when sailing under a leeway angle.

2.2 (a) Prediction of K_T and K_Q using w_{AX}

In straight-sailing conditions the observed change in thrust and torque of the propeller from open-water conditions is generally attributed to a change in w_{AX}. The value of the wake fraction is however not determined by calculating the mean axial velocity from a measured or calculated wakefield, but using the so-called thrust-identity method.

For this method the measured value of K_T is matched with the K_T curve in the open-water diagram. The difference between the found advance coefficient (J) and the apparent advance coefficient (J_v) gives the wake fraction using Equation 7. The wake fraction obtained using this method is called the effective or Taylor wake fraction (w_T). The difference in measured torque compared to the open-water diagram at J, shown in Equation 8, results in a deviation of the relative rotative efficiency (η_R) from 1. This procedure is illustrated in Figure 4.

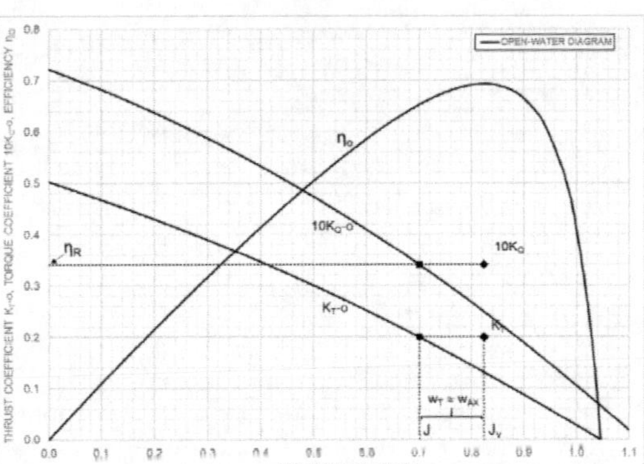

Figure 4: Example of the determination of w_T and η_R using the thrust-identity method

$$w_T = 1 - \frac{J}{Jv} \quad (7)$$

$$\eta_R = \frac{K_{Q-o}}{K_Q} \quad (8)$$

In the absence of significant pre-swirl, w_T is generally close to w_{AX} and could be used to estimate K_T and K_Q in behind condition.

2.2 (b) Prediction of K_T and K_Q using w_{AX} in case of pre-swirl

In case of significant pre-swirl, the measured coefficients (J, K_T, K_Q) no longer coincide with the coefficients in the open-water diagram (J_{-O}, $K_{T\text{-}O}$, $K_{Q\text{-}O}$) when only w_{AX} is considered. An additional effect is present due to the change in rotation rate experienced by the propeller due to the pre-swirl.

The coefficients do however match in case of uniform flow rotation in the wakefield when the thrust and torque would have been divided by the sum of the propeller rotation rate (n_p) and the pre-swirl rotation rate ($n_{pre\text{-}swirl}$) instead of just n_p. This transformation is shown in Equation 9, 10 and 11. The measured coefficients (J, K_T, K_Q) are calculated based on n_p as is generally done.

$$J_{-o} = \frac{V_A}{(n_p + n_{pre-swirl})D} = J\frac{n_p}{(n_p + n_{pre-swirl})} \quad (9)$$

$$K_{T-o} = \frac{T}{\rho\,(n_p + n_{pre-swirl})^2 D^4} = K_T\frac{{n_p}^2}{(n_p + n_{pre-swirl})^2} \quad (10)$$

$$K_{Q-o} = \frac{Q}{\rho\,(n_p + n_{pre-swirl})^2 D^5} = K_Q\frac{{n_p}^2}{(n_p + n_{pre-swirl})^2} \quad (11)$$

When w_{AX} and $n_{pre\text{-}swirl}$ are known, the same equations can be used to predict the expected K_T and K_Q coefficients The procedure is given below and illustrated in Figure 5:

- Transform the open-water curves using the equations below. (Dashed lines in Figure 5)

  $$J^* = J_{-o}\frac{(n_p + n_{pre-swirl})}{n_p} \quad (12)$$

  $$K_T^* = K_{T-o}\frac{(n_p + n_{pre-swirl})^2}{{n_p}^2} \quad (13)$$

  $$K_Q^* = K_{Q-o}\frac{(n_p + n_{pre-swirl})^2}{{n_p}^2} \quad (14)$$

- Obtain the K_T and K_Q values along the transformed curve at J_{AX}.

  $$J_{AX} = J_V\,(1 - w_{AX}) \quad (15)$$

- Obtain w_T and η_R using the thrust-identity method and Equation (7) and (8)

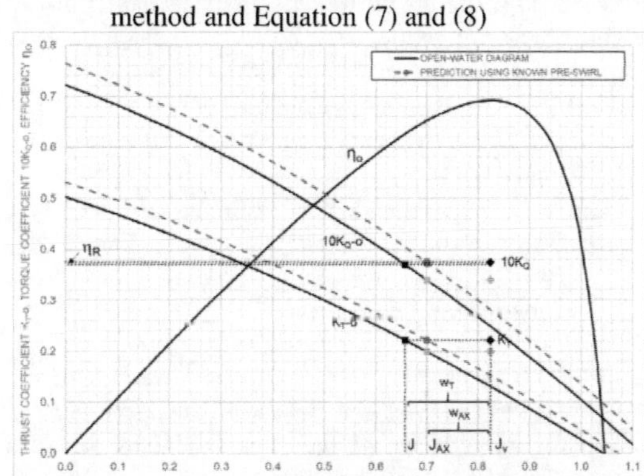

Figure 5: Example of the prediction of K_T, K_Q, w_T and η_R using a known axial wake fraction and pre-swirl rotation rate

2.2 (c) Determination of the propulsive efficiency

The propulsive efficiency is determined and reported based on Equation 16. The values of K_T and K_Q are obtained using the procedure described in paragraph 2.2 (b). The value of $K_{Q\text{-}o}$ is interpolated at J. Instead of w_T, w_{AX} is used to divide the propulsive efficiency (η_D) more realistically in the hull efficiency (η_H) and open-water efficiency (η_O) when a contribution of pre-swirl is present. The thrust deduction is assumed constant over the leeway angels in the current analysis [3]. A non-constant thrust deduction would affect the trend in propulsive efficiency.

$$\eta_D = \frac{P_E}{P_D} = \frac{R\,V_s}{M_p\,n_p} = \underbrace{\left(\frac{K_T}{K_{Q,o}}\frac{J_v\,(1-w_{AX})}{2\,\pi}\right)}_{\eta_o}\underbrace{\left(\frac{K_{Q,o}}{K_Q}\right)}_{\eta_R}\underbrace{\left(\frac{1-t}{1-w_{AX}}\right)}_{\eta_H} \quad (16)$$

4. Prediction compared to captive model tests

This paragraph describes the flow observations for a twin-screw vessel and three single screw vessels with a gondola aft ship form. The value of w_T obtained with the prediction method described in paragraph 2.2 (b) is compared to results from captive model tests over leeway angels. All the results are obtained at model scale.

4.1 TWIN-SCREW VESSEL – 5415M

The effects of leeway on the propeller performance for twin-screw vessels was analysed based on 5415M. This is a preliminary design of a navy surface with a sonar dome and transom stern. The design has been extensively used for validation of viscous flow calculations in various workshops.

4.1 (a) Viscous flow calculation results

Viscous flow calculations at various leeway angles were used to investigate the flow pattern around the hull and change in wakefield. An illustration of the flow pattern is given in Figure 6. In the figure the relative axial velocity is shown on slices along the hull and in the propeller plane. The vortices forming around the hull are indicated in grey. The main vortices observed in the flow are:

- At β=-4° the vortex from the keel line at the bow just aft of the sonar dome reaches the propeller plane.
- At β=-12° the vortex from the aft keel line passes through the propeller plane. The vortex forms at the stabilizer fin and moves over the centre skeg.

The effect of these vortices can also be observed in the trend of w_{AX} and $n_{pre\text{-}swirl}$' obtained from the nominal field given in Figure 7. It can be seen that w_{AX} increases at a leeway angle of -4 deg when the vortex passes through propeller plane. A significant increase in pre-swirl is also visible due to the vortices. The general trend of w_{AX} is mainly caused by the decreasing u-velocity due to leeway angle described in paragraph 2.1.

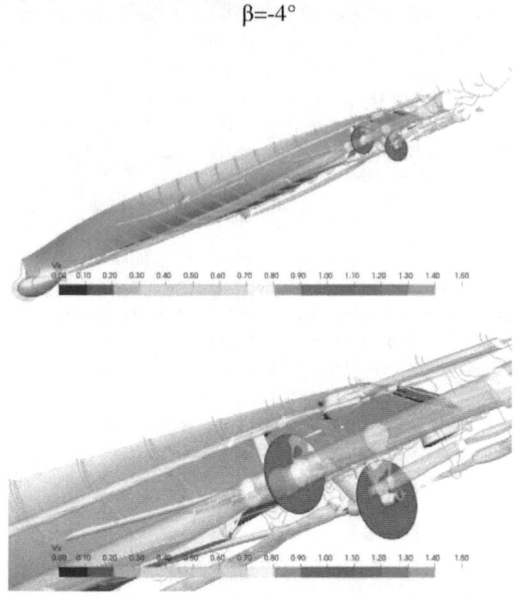

Figure 6: Flow towards the propeller plane for the 5415M naval combatant shown at two leeway angles.

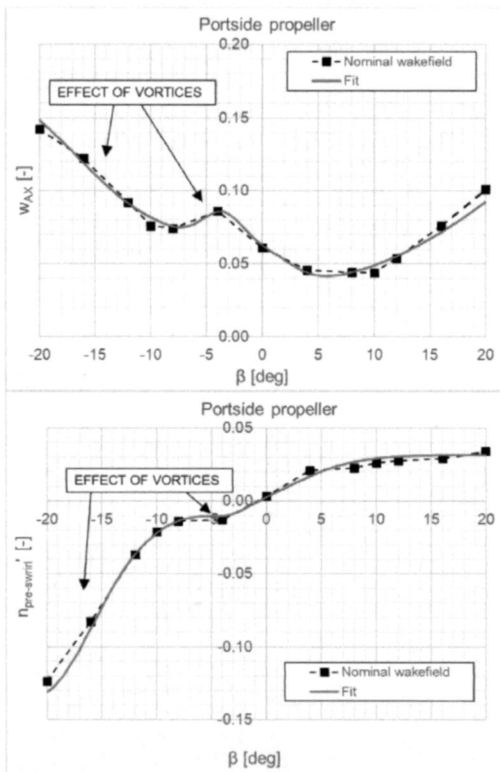

Figure 7: Axial wake fraction (w_{AX}) and non-dimensionless pre-swirl rotation rate over leeway angles for the port side propeller of the 5415M, showing the effect of vortices.

4.1 (b) Comparison with captive model tests

Two captive test campaigns were available to compare against the prediction model. It should be noted that the pitch of the used propeller was different.

Using the obtained fits of w_{AX} and $n_{pre\text{-}swirl}$' the value of w_T was determined using the theory described in paragraph 2.2 (b). In Figure 8 it is observed that the trend of the captive model tests is reasonably well predicted. An interesting observation is that the trend of w_T differs very much from the trend in w_{AX} shown on the top in Figure 7. The difference indicates the importance of the pre-swirl for vessels sailing at a leeway angle. Due to the pre-rotation the value of w_T can depend on the propeller rotation rate and pitch.

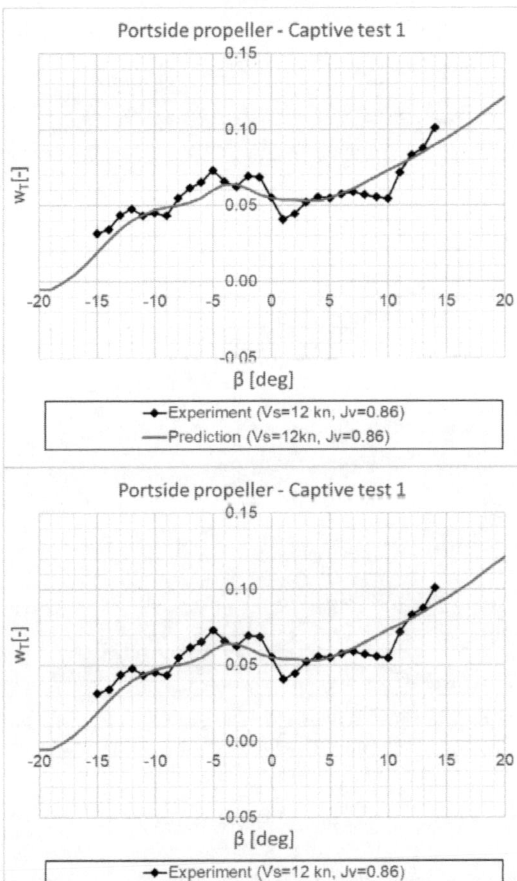

Figure 8: Comparison of the predicted w_T and measured values for the two model test campaigns for the 5415M.

4.2 SINGLE-SCREW VESSELS

The effects of leeway on the propeller performance for single-screw vessels with a gondola stern were studied for three vessel types:

- Container ship (KCS)
- Very Large Crude Carrier (KVLCC2)
- Coaster (WHC)

The first two designs are extensively used for validation of viscous flow calculations in various workshops and for this reason captive test data is available. The Wind Hybrid Coaster (WHC) was subject to model tests at MARIN during wind-propulsion related research.

4.2 (a) Viscous flow calculation results

Viscous flow calculations at various leeway angles where again used to investigate the flow pattern towards the propeller. The results for the container ship and VLCC are shown in Figure 11. The relative axial velocity is shown on slices along the hull and in the propeller plane. The vortices forming around the hull are indicated in grey and streamlines leading to the propeller plane in black. The main flow observations are as follows:

- The axial velocity is reduced “downstream” of the gondola (see also the wakefield in Figure 3).
- A vortex forms due to the flow over the gondola, inducing pre-swirl

A strong interaction between the propeller suction and the flow towards the propeller is observed for these vessel types, which was not observed for the twin-screw vessel. The lower velocities in one half of the wake field lead to an increase in propeller suction. This in turn leads to a stronger flow around the gondola and increased pre-swirl.

For this reason the effective wakefield was used to determine w_{AX} and $n_{pre\text{-}swirl}$' instead of the nominal wakefield. This wakefield is obtained by using a propeller model in the viscous flow calculation and subtracting the propeller induced velocities from the wakefield. This still leaves the general modification to the flow due to the propeller action.

The results are given in Figure 9, where it should be noted that the fit for $n_{pre\text{-}swirl}$' had to be tuned towards the captive model test data. Most likely this was necessary due to a well-known sensitivity of the resulting vortices to the turbulence modelling. A sensitivity study for the KCS at β=8° showed that a variation in the turbulence model could increase the $n_{pre\text{-}swirl}$' value with 60% due to slight changes in the location relative to the propeller plane and strength of the gondola vortex. The experimental data are therefore valuable for obtaining the realistic trends.

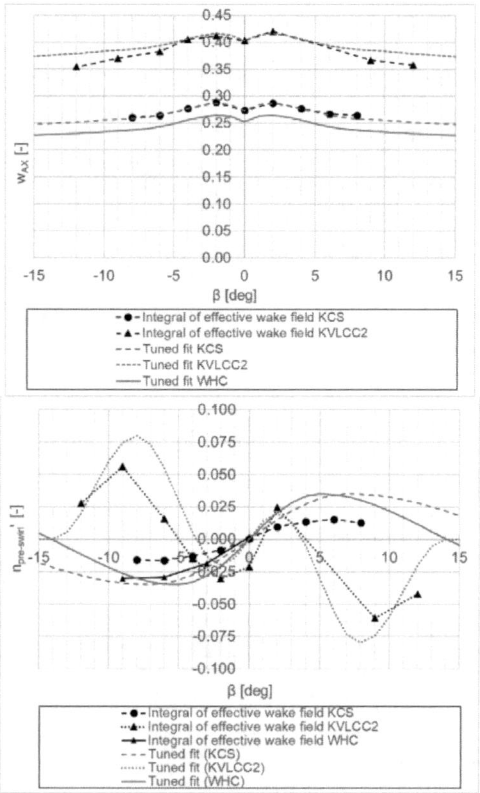

Figure 9: w_{AX} and $n_{pre\text{-}swirl}$' obtained from the effective wakefields calculated for the single-screw vessels

It is observed that the trend for the pre-swirl of the KVLCC2 differs from the other two vessels studied for the larger leeway angels. In a comparison of the wakefields of the KCS and KVLCC2 shown in Figure 10 it is observed that this is related to the location of the vortex relative the propeller plane. On the right side the rotation induced by the vortex from the gondola is visible, creating much stronger downward velocities near the hub for the KVLCC2. For the KCS the rotation in the other direction near the bottom is more prominent.

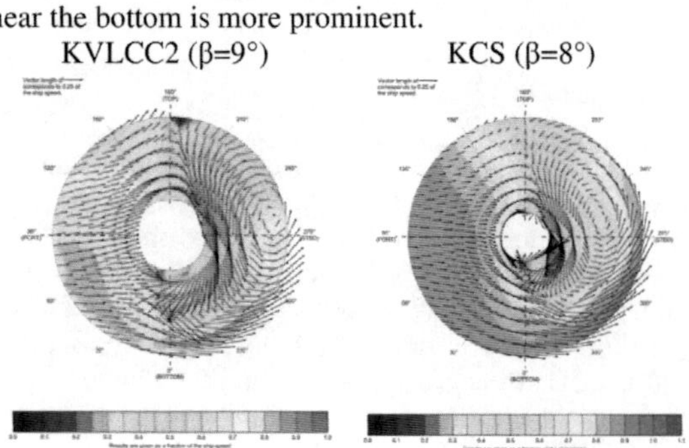

Figure 10: Comparison between the effective wakefield for the VLCC (KVLCC2) and the container ship (KCS).

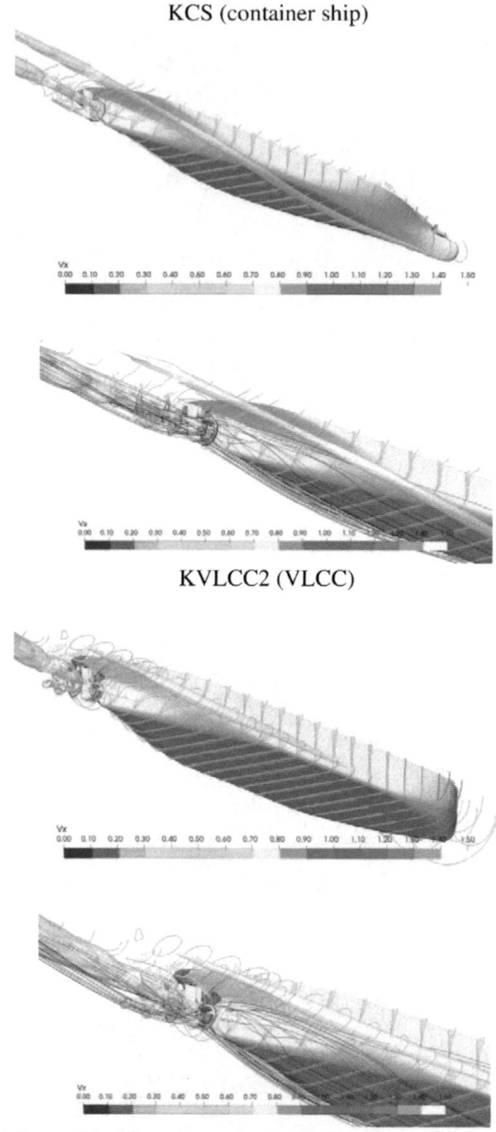

Figure 11: Flow towards the propeller plane for the KCS container ship shown at β=8° (top) and the KVLCC2 at β=9° (bottom).

4.2 (b) Comparison with captive model tests

The comparison of the tuned fit with the captive test results for the KCS container ship is shown in Figure 12. Only the ship model used in test campaign 2 included bilge keels. The vortices from the bilge keels reach the propeller plane around β=8°, where an additional increase in w_T is observed relative to captive test campaign 1. This is likely due to these bilge keels. It is observed that the general trend in the asymmetry in w_T at positive and negative leeway angles is well captured, this asymmetry is also reported in literature [3]. It is caused by the pre-swirl in the wake field in combination with the rotation direction of the propeller.

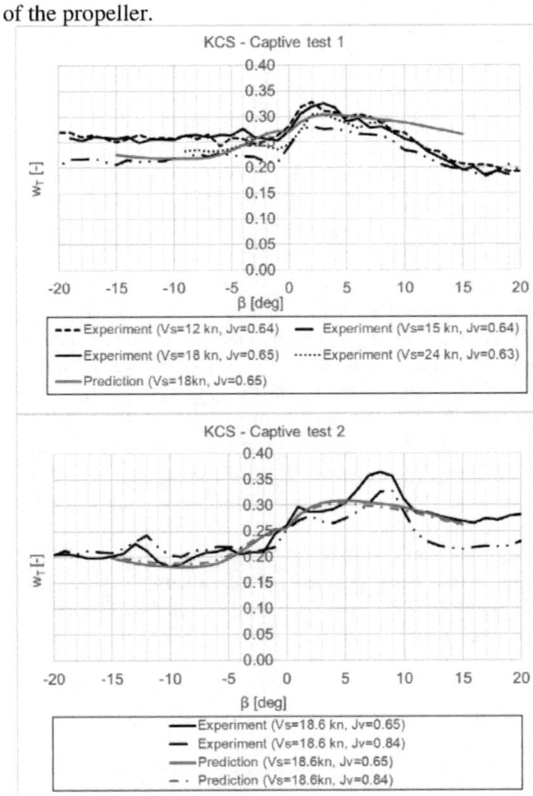

Figure 12: Comparison of the predicted w_T and measured values for the two model test campaigns on the KCS container ship. The second model test campaign included bilge keels.

For the KVLCC2 data is available from two captive test campaigns for the SIMMAN workshop [4]. The first test series was performed at HMRI and the second at NMRI for this workshop. The results are compared with the prediction in Figure 13. It is observed that again the asymmetry and general trend is captured by the prediction model. The asymmetry is over-predicted for the negative leeway angles compared to the experiments.

The trend differs with respect to the KCS container ship shown before. This is related to the different behaviour of the pre-swirl rotation in the wakefield over leeway angles as discussed in paragraph 4.2 (a) and shown in Figure 10.

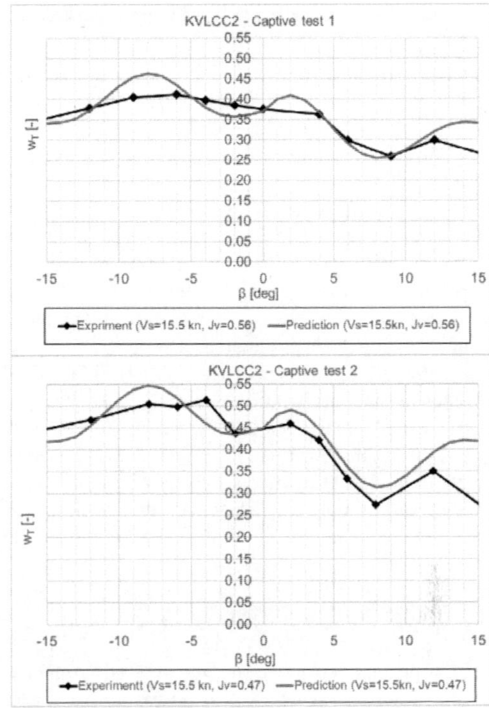

Figure 13: Comparison of the predicted w_T and measured values for the two model test campaigns on the KVLCC2.

For the WHC design tests for only a limited number of leeway angles were considered during the captive tests [5]. The results are compared with the prediction in Figure 14 and show that the general trend is followed. The used trend for w_{AX} was based on the fit for the KCS and KVLCC2 as the axial wake fraction obtained from the available viscous flow results was not reliable.

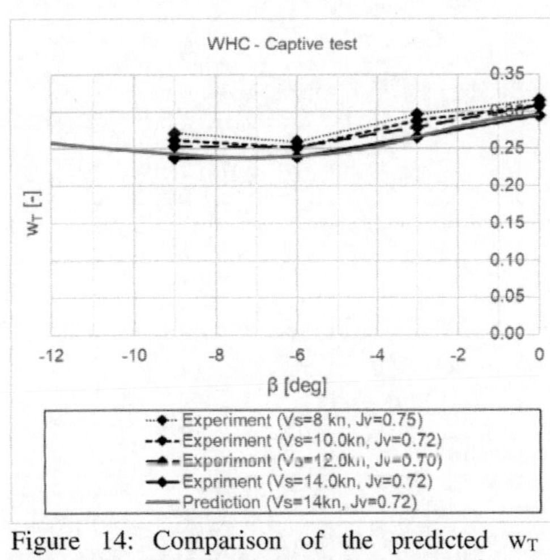

Figure 14: Comparison of the predicted w_T and measured values for the model test campaign on the WHC.

5. PROPELLER PERFORMANCE TRENDS

Based on the obtained fit of w_{AX} and $n_{pre\text{-}swirl}$' the trends in delivered propeller power and propulsive efficiency can be shown using the proposed prediction model. The results are based on an assumed constant thrust deduction as discussed earlier. The wake fraction is based on model scale results, for full scale mainly the axial wake fraction is much lower for the single screw vessels. The trends over leeway angles are likely to be similar between model and full scale. The results are given along lines of constant propeller rotation rate and constant ship speed as the percentage change of propulsive efficiency (η_D) and delivered propeller power (P_D) from the straight sailing value. This is a simplification. In reality the propeller rotation rate will change due to the described effect of the wake and the required thrust that varies with wind speed and angle in a wind assistance scenario. These effects are covered in follow-up Performance Prediction Programs (PPP) at MARIN but are not discussed in the present paper.

5.1 TWIN-SCREW VESSEL – 5415M

First the approximated trends for the studied twin-screw vessel are shown in Figure 15. It is observed that for low rotation rates (high J_v) the effects of the leeway increases. Overall the differences from straight-sailing conditions are small for this twin-screw vessel at small leeway angles. For larger leeway angles larger changes in delivered propeller power are found. These effects are less important for the wind-assisted scenario but do influence the behaviour during manoeuvres such as a turning circle and affect the operational point within the engine envelope during these manoeuvres.

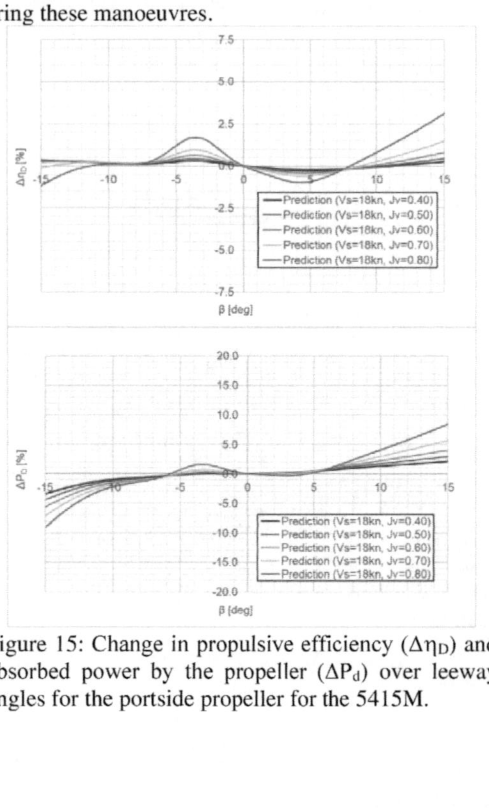

Figure 15: Change in propulsive efficiency ($\Delta\eta_D$) and absorbed power by the propeller (ΔP_d) over leeway angles for the portside propeller for the 5415M.

5.2 SINGLE-SCREW VESSELS

Also for the three single screw vessels the trends in absorbed propeller power and propulsive efficiency are studied relative to straight-sailing condition. The results for constant rotation rate and ship speed are shown in Figure 16, Figure 17 and Figure 18. The largest apparent advance ratios (J_v) shown are for the propeller operating at J values near the maximum open water efficiency. Such high J_V are likely also import for wind assisted ships.

It is observed that also for all the single screw vessels the change in propeller performance due to leeway angle increases with apparent advance ratio (J_v). For a higher J_v value at the same ship speed, the propeller rotation rate is lower. The relative effect of the pre-swirl is therefore larger than for a faster rotating propeller. For wind assisted ships the propeller rotation rate when maintaining constant ship speed will decrease, as part of the thrust is generated by the wind propulsion device instead of the propeller. This will lead to an increased effects of the leeway angle on the propeller. However, the increase of J_v in straight sailing condition generally increase the propulsive efficiency further than only the effects of leeway considered here.

Between the single screw ship types studied the highest sensitivity of the propeller performance to leeway effects is observed for the coaster design (WHC). The container ship (KCS) has a lower sensitivity to leeway but a similar trend as for the coaster. The VLCC (KVLCC2) shows the least sensitivity to leeway angles, although the trend differs from the other two ship types studied due to the differences in the gondola vortex described earlier.

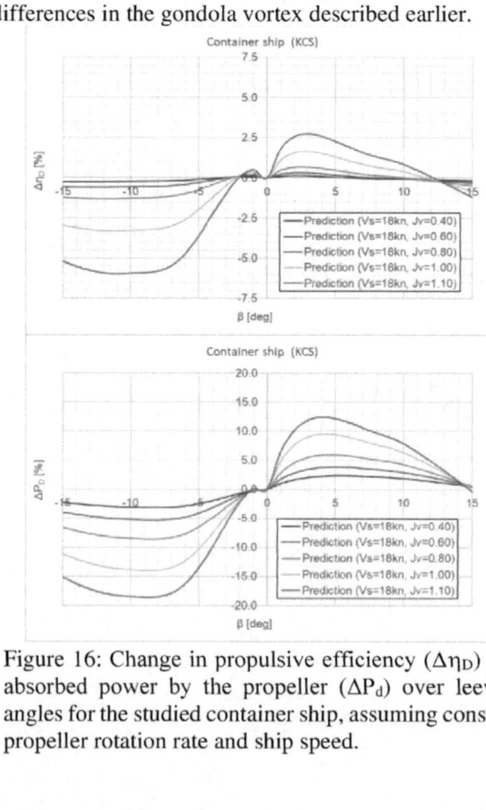

Figure 16: Change in propulsive efficiency ($\Delta\eta_D$) and absorbed power by the propeller (ΔP_d) over leeway angles for the studied container ship, assuming constant propeller rotation rate and ship speed.

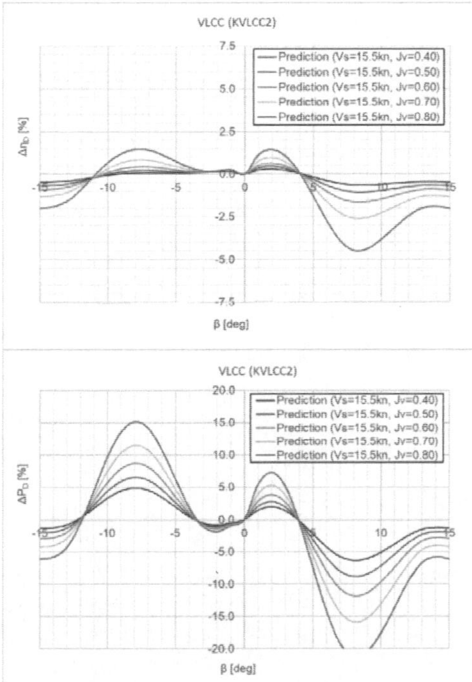

Figure 17: Change in propulsive efficiency ($\Delta\eta_D$) and absorbed power by the propeller (ΔP_d) over leeway angles for the studied VLCC, assuming constant propeller rotation rate and ship speed.

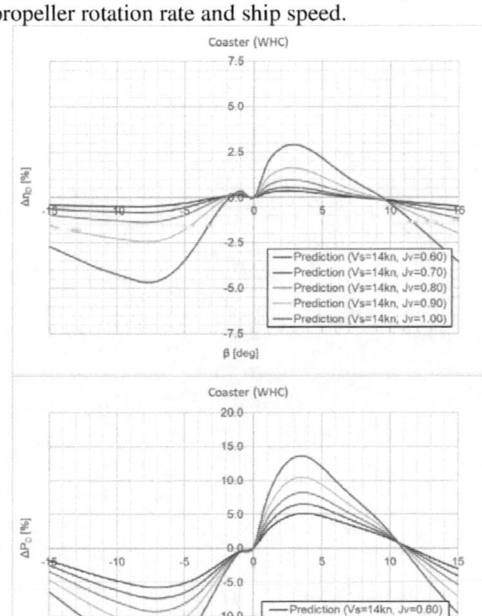

Figure 18: Change in propulsive efficiency ($\Delta\eta_D$) and absorbed power by the propeller (ΔP_d) over leeway angles for the studied coaster, assuming constant propeller rotation rate and ship speed.

7. CONCLUSIONS

A design stage prediction method is proposed to predict the effects of leeway on the propeller thrust and torque using a fit of the axial wake fraction and pre-swirl over leeway angles in combination with the open-water diagram. This pre-swirl is present due to vortices forming around the hull when sailing at a leeway angle. The average side velocity that is also present in the wakefield has a smaller effect on the thrust and torque and could be neglected in the prediction. However, it can contribute to other effects such as bearing loads, rudder forces and cavitation. This was not considered during this study.

The method results in a prediction of the propeller thrust and torque coefficient which can be related to the Taylor wake fraction (w_T) and relative rotative efficiency (η_R). These interaction coefficients are often input for empirical prediction methods together with the open-water diagram and therefore of interest. It should be noted that due to the pre-swirl present in the wake these coefficient are not constant for changing propeller rotation rate and propeller pitch as normally might be assumed.

The predicted value of w_T is compared to the results of captive model tests of a twin-screw vessel and three single-screw vessels with a gondola stern. The axial w_{AX} and $n_{pre\text{-}swirl}$', which is input to the prediction, was obtained from integration of the wakefields available from viscous flow calculations. For the twin-screw vessel a good correlation was obtained using the calculated nominal wakefield. The observed $n_{pre\text{-}swirl}$' was mainly related to a vortex from the bow and stern passing the propeller plane.

For the single-screw vessels the effective wakefield had to be used to obtain the value of $n_{pre\text{-}swirl}$' that is expected based on the captive model tests. This is related to the observed strong interaction between the propeller suction and gondola flow. Still an under prediction of the pre-swirl was observed, most likely due to deficiencies of the turbulence modelling for the prediction of vortices. This could be corrected for using the captive model test results.

Finally, the prediction method is used to approximate the change in absorbed propeller power and propulsive efficiency for the vessel types studied over leeway angles. Only the effect over lines of constant ship speed and propeller rotation rate are shown in this study. The thrust deduction was assumed constant over leeway angles and model scale wake results were used.

The most significant effects on absorbed power and propulsive efficiency are observed for the single-screw vessels. The coaster design shows the largest sensitivity and smaller changes are observed for the studied container ship and VLCC design. The effect of leeway increases for a lower propeller rotation rate at the same ship speed. This is caused by the pre-swirl in the wakefield which stays similar and therefore relatively increases compared to the propeller rotation rate. The variation over leeway angles on the delivered propeller power is in the order of 10% and on the propulsive efficiency in the order of 2%. The change in power and efficiency will differ when also the propeller rotation rate at self-propulsion point is taken into account for the wind assisted scenario. This effect is covered in follow-up Performance Prediction Programs (PPP) calculations at MARIN but are not discussed in the present paper.

8. ACKNOWLEDGEMENTS

This research is partly funded by the Dutch Ministry of Economic Affairs.

9. REFERENCES

1. VAN TERWISGA T., 'On the working principles of Energy Saving Devices', *Third International Symposium on Marine Propulsors*, 2013.

2. SCHUILING B., 'The Design and Numerical Demonstration of a New Energy Saving Device', *Numerical Towing Tank Symposium*, 2013.

3. KOSE K., 'On a new mathematical model of maneuvering motions of a ship and its applications', *International Shipbuilding Progress vol.29 no.336 pp. 205-220*, 1982

4. SIMMAN, 'Workshop proceedings', *Workshop on verification and Validation of ship manoeuvring simulation methods* , 2014.

5. EGGERS R., 'Operational Performance of Wind Assisted Ships', *Hiper conference*, 2016

9. AUTHORS BIOGRAPHY

Joost Schot holds the position of Project Manager - Ships at the Maritime Research Institute Netherlands. He is responsible for the application of CFD and other prediction methods within the transport and shipping market segment.

Rogier Eggers holds the position of Senior Project Manager and Knowledge co-ordinator of Ships Manoeuvring at the Maritime Research Institute Netherlands. His main business is to conduct model tests and simulations and to help clients regarding manoeuvring and seakeeping of ships. Since 2011 his work also includes research regarding Wind Assisted Ship Propulsion.

PERFORMANCE VERIFICATION OF A RECENT ROTOR SAIL INSTALLATION

V. Paakkari and **O. Valkeisenmäki,** Norsepower Ltd, Finland

SUMMARY

Auxiliary wind propulsion offers viable solution to reduce the emissions of global shipping. While it is very traditional means of propulsion, it is relatively new to modern commercial shipping. It is then highly important to provide the industry with independent and verified data of its performance. This paper reviews measurement campaigns conducted onbaord cruise ferry Viking Grace, which was retrofitted with a Rotor Sail in April 2018. Four different approaches to the performance verification were applied and the results show good agreement between results from different methods. These results are also seen to be in-line with the original target of the project.

1. INTRODUCTION

1.1 REDUCING EMISSIONS IN GLOBAL SHIPPING

The challenge of reducing GHG emissions of shipping and eventually making the industry carbon neutral will require many new technologies to be developed. Although it is not likely that any one technology will become a silver bullet to solve the problem on its own, it is probable that wind propulsion will frequently be included in the mix of used solutions. As a primary renewable source of energy, wind propulsion does not result in increased demand of overall energy production capacity. An increased number of modern wind propulsion technologies have also a very high technology readiness level, and some products have already reached full market readiness. Rotor Sails are a market ready solution to provide the shipping industry with auxiliary wind propulsion. In this paper, the focus is in performance verification of the technology on a cruise ferry MS Viking Grace, which was equipped with a Rotor Sail in 2018.

1.2 ROTOR SAIL TECHNOLOGY

Rotor Sails are a modernized version of Flettner rotors, mechanical sails utilizing a principle called Magnus effect. The basic principle is, that a rotating object induces pressure difference between its sides when it is exposed to a flow (wind). Essentially Flettner rotors are lift producing devices like airplane wings or traditional sails of a sailboat. However, the lift production capability of a Rotor Sail, thanks to the Magnus effect, is several times higher when compared to a conventional airplane wing. This makes Rotor Sails a very interesting source of auxiliary wind propulsion on ships. Figure 1 shows a schematic situation, where a round, rotating object is exposed to wind. When the object is rotating, a small boundary layer is formed around it. This layer of air follows the rotating surface of the object. On one side of the cylinder this boundary layer flows against the wind flow decelerating it, and on the other side both wind and boundary layer flow in same direction and thus the boundary layer accelerates the wind flow.

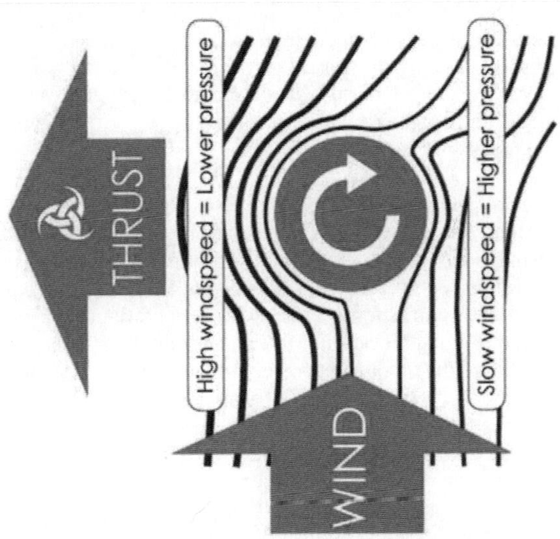

Figure 1: Schematic illustration of the Magnus effect. Spinning round object is exposed to wind depicted with a red arrow.

In the convention of the Figure 1, the wind speed on the left-hand side is accelerated and on the right-hand side it is decelerated. This change in wind speeds results in pressure difference according to so called Bernoulli principle; the increased speed leads to lower pressure, and decreased speed leads to higher pressure. The pressure difference between the sides results in a force vector, which is perpendicular to the wind (depicted with red "Thrust" arrow in Figure 1).

Norsepower offers three standardised Rotor Sail models with heights of 18, 24, and 30 meters and corresponding rotor diameters of 3, 4 and 5 meters. A Rotor Sail consists of two main components:

- A rotating cylinder (the rotor), made of a laminated glass fibre and carbon fibre composite.
- A steel support tower inside the rotor extending up to circa two thirds of the rotor height.

The Rotor Sails transfer the forces generated by the cylinder surface to the ships hull via vessel-tailored foundations. The Rotor Sails are installed on the foundations with a bolt connection. When the installation of the foundations has been completed, the Rotor Sails can be lifted on the vessel and connected to the foundations within some hours. This enables fast installation of the actual sail, and it can typically be done during a normal port call. To maintain the rotational movement, Rotor Sails are equipped with a variable speed drive and an electric motor, which is powered by the ships low-voltage power grid. The operation of Rotor Sails is fully automated for optimising the energy saving performance and they don't increase the workload of the crew.

1.3 PERFORMANCE VERIFICATON

The Magnus effect and the Flettner rotors have been known over a century, but the commercial implementation in shipping has not taken place until more recently. Because of this, Flettner rotors can be viewed as a new product entering the market. To gain market acceptance, there are several hurdles that need to be tackled, most importantly the verification of the thrust performance and the fuel savings potential.

The objective of this paper is to contribute to this challenge by presenting the results on one of the recent installations of a Rotor Sail on a cruise vessel. In April 2018, Norsepower's model 24x4 Rotor Sail was installed onboard a cruise ferry Viking Grace operating between Turku (FI) and Stockholm (SWE). During the first year of operation, the propulsion power reduction and fuel saving performance of the Rotor Sail was investigated by three independent parties. In addition, the data logging of thrust performance was done by Norsepower.

In chapter two, these measurement campaigns and their methodologies are described and in chapter three, the results of them are presented. The final chapter discusses the findings and provides a short outlook on the future potential of Rotor Sails.

2. METHOD

2.1 ROTOR SAIL ON VIKING GRACE

The Rotor Sail was installed onboard Viking Grace in April 2018. Vessel operates on a daily basis between Turku and Stockholm. The vessel was one of the first LNG-powered passenger vessels, and it was the first passenger vessel to utilize a Flettner rotor as an auxiliary propulsion system. The sail is 24 meters high and it has a diameter of 4 meters. After the installation, the propulsion power reduction due to the Rotor Sail was analysed by three independent parties: NAPA, ABB and Chalmers University. In addition, Norsepower made its own thrust measurements. All approaches were slightly different and used different methodologies to determine the fuel savings. The parties were also kept separate from each other so, that the analyses were as independent as possible. In the following, a brief description of each campaign is given.

2.2 NAPA METHODOLOGY

NAPA performance analysis campaign [1] was conducted between October 2018 until January 2019. When comparing the performed measurements based on propulsion power, the approach taken by NAPA was the most straight-forward. Two separate phases can be distinguished.

In the first phase, a measurement environment as controlled as possible was created. First, this meant defining measurement legs on vessel's route; legs were defined so, that the conditions would be as constant as possible over the leg (i.e. open water legs with constant speed). Secondly, the Rotor Sail automation was configured so, that half of the time when crossing these legs, the Rotor Sail would be disabled and half of the time it would be in operation. This way, two data sets with clear categorial variable (sail enabled / disabled) could be defined.

In the second phase, these data sets were further post-processed to control other variables influencing the propulsion power. Based on the data sets, a statistical regression model was built to determine the influence on the vessel's propulsion. The representativeness of the statistical regression model can be described by the R-squared value of the model, which was reported by NAPA being 93%. This means, that the model explains the collected data very well.

While the use of pre-defined measurement legs enables better measurement conditions, the analysis cannot be directly applied to areas outside of the measurement legs.

2.3 ABB METHODOLOGY

ABB had the longest measurement campaign [2]. The company has a propulsion performance monitoring system installed onboard Viking Grace, so they were able to use also data which was collected before the Rotor Sail was installed. The data collection with Rotor Sail installed took place between July 2018 and March 2019.

From the operational data with various input signals (trim, draft, list, ...), a so-called propulsion power breakdown model was created. This model was compiled by analysing the effects of these variables to the propulsion power requirement. Then, when installing new energy saving equipment, the model can be recalibrated based on the data analysis to include also additional equipment. The analysis searches for best possible mapping between signal inputs and propulsion power requirement. The analysis has a foundation on theoretical considerations of qualitative behaviour of different phenomena contributing to the propulsion power requirement, but the quantitative weights of different phenomena are determined by data analysis of the propulsion power measurement.

The advantage of this type of analysis is, that it considers the actual operational variables contributing to the propulsion power, while not requiring the Rotor Sail to be stopped frequently, which would naturally decrease the overall savings during the measurement campaign.

2.4 CHALMERS UNIVERSITY METHODOLOGY

The performance analysis by Chalmers University [3] was based on their in-house developed propulsion power model called ShipCLEAN. The simulation model was validated regarding the ship and the Rotor Sail model with the measurement data logged onboard of MS Viking Grace from September 2018 until the end of January 2019. The predictions by the simulation were compared against the measurements, and they were stated to have sufficient accuracy for the study. The simulation prediction of the annual savings potential was based on environmental data collected onboard Viking Grace during 2018. This means, that the savings reported in the study are not actually savings produced by Rotor Sail, but actually measurement-verified simulation runs using the data of 2018 i.e. estimating, what would have been the savings if the Rotor Sail had been in operation the whole year.

2.5 STRAIN MEASUREMENT

All of the measurements presented so far rely on direct measurement of propulsion power and then analysing the contribution of Rotor Sail by different methods. To supplement these measurements, Norsepower conducted a direct strain measurement on the Rotor Sail. This method yields information on forces acting on Rotor Sail, but it does not readily tell about the propulsion power or fuel savings. In order to calculate these, a conversion formula is required between force and propulsion power equivalent. The conversion [4] is presented in equation 1.

$$P_{RS} = \frac{T_{RS}V}{\eta} \quad (1)$$

Where T_{RS} is thrust produced by the Rotor Sail, V is the vessel speed and η is total propulsion efficiency defined by

$$\eta = \frac{P_E}{P_B} \quad (2)$$

In which, P_E is effective power of the vessel and P_B is the brake power.

The strain gauges were installed on the internal support tower of the Rotor Sail, and the measurement was calibrated onshore before the Rotor Sail was installed onboard.

3. RESULTS

Due to different approaches taken, the daily route of Viking Grace was divided to "Open Sea legs" and "Archipelago legs" (see chapter 2.2 for further information).

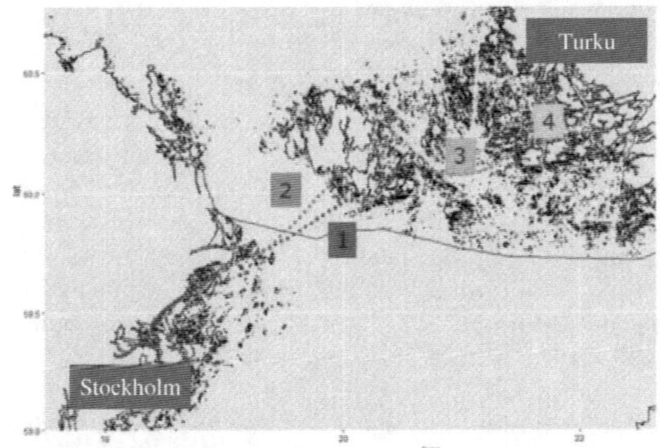

Figure 2: Daily route of Viking Grace and the defined measurement legs (numbered)

This division was done for measurement purposes, but it is evident that the external conditions and operational profile are quite different in these different groups. Regarding the operational environment, especially on legs 1 and 2 in Figure 2, the wind conditions are less affected by the islands compared to the legs, which are located deep in the archipelago. As a result, it is seen that the annual average wind speeds are somewhat higher in this area than elsewhere along the route [5]. Perhaps even more importantly, the average speed of Viking Grace is much higher on the "Open Sea legs". On these legs, the vessel is sailing most of the time over 20 knots, whereas on the "Archipelago legs", the average speed is around 8-10 knots. These two factors contribute to significant differences in the savings reported on different legs.

3.1 OPEN SEA LEGS

The results of measurements on Open Sea legs are presented in table 1. The analysis conducted by Chalmers does not include division between different leg types, so it is omitted from tables 1 and 2. From the table it is seen, that the highest estimate for propulsion power savings is given by ABB, while strain measurement by Norsepower is the most conservative. However, it is also seen that the results have relatively narrow spread, which increases the confidence on the measurements.

	ABB	NAPA	Strain
Analysis period	**Jul 18 – Mar 19**	**Oct 18 – Jan 19**	**Nov 18 – Mar 19**
Est. average net saving when Rotor Sail enabled [kW]	341	282	257
Equals annual fuel saving (on open sea legs only) [ton]	109	90	82
Average forward thrust when Rotor Sail enabled [kN]	N/A	N/A	12

Table 1. Performance measurement on "Open Sea Legs"

3.2 ARCHIPELAGO LEGS

The results of measurements on Archipelago legs are presented in table 2. As can be seen, even if the average measured thrust forces are very close to each other on archipelago and open sea legs, the propulsion power savings are much higher in the latter case. This can be explained through equation (1), where increasing the vessel's speed (which is roughly two-fold in this case) increases the propulsion power equivalent i.e. the savings of the Rotor Sails.

Again, ABB result are higher than the results from strain measurement. There are two possible reasons for this. The first one is related to the calculation method of how the strain is converted into propulsion power equivalent.

	ABB	Strain
Analysis period	**Jul 18 – Mar 19**	**Nov 18 – Mar 19**
Estimated average net saving when Rotor Sail enabled [kW]	153	99
Equals annual fuel saving (on archipelago legs only) [ton]	122	79
Average forward thrust when Rotor Sail enabled [kN]	N/A	11

Table 2. Performance measurement on "Archipelago Legs"

From Equation (1) it is seen, that besides thrust force and speed measurement, also the total efficiency of the vessel affects the measurement. For the conversion, fixed efficiency is used, and underestimating the efficiency would also lead to the differences between different analysis methods.

Another possible explanation would be errors in calibration, which was done before installation. In this case, differences in foundation's mechanical properties (ship vs shore) could lead to a scale error.

3.2 LONG TERM SAVING POTENTIAL

Based on the measurements, long-term estimate for fuel savings potential was deducted separately from each campaign. It is important to note that of these estimates, only the estimate by Norsepower is genuine long-term estimate utilizing information of long-term weather data. Others are in sense extrapolations of measurement period over the annual time frame.

	ABB	NAPA	Chalmers	NP
Estimated long-term average net saving based on all data [kW]	(207)	(282)	(287)	211
Equals annual fuel saving (on all legs of Viking Grace route) [ton]	(231)	(315)	(320)	236

Table 3. Long term estimates of the savings potential based on measured savings.

The following remarks need to be made for long term estimates of external parties:

> ABB: Long term potential is assumed to be the same as average saving during the measurement campaign when Rotor Sail was enabled. Due to the enable/disable measurement campaign, this is likely to be an underestimate (half of the open sea legs, with highest savings, are absent).
>
> NAPA: Long term potential is assumed to be same as average savings during the measurement campaign. Due to neglecting the archipelago legs (with lowest savings), this is likely to be an overestimate.
>
> Chalmers: Long term potential is assumed to be same as simulated results with environmental data on board during 2018

The original target for the fuel savings on Viking Grace was up to 300 tons of LNG per year. With the aforementioned notions in mind it can be concluded, that the project has shown to be successful in reaching its annual savings target very closely, and that the original target falls within the range of measurements.

4. DISCUSSION

In the previous chapter it has been presented, that Rotor Sail onboard Viking Grace has closely met its targets in terms of fuel saving potential. As a further notice, the overall feedback from cruise customers has been positive, and introduction of Rotor Sail has been assessed to improve the company image of Viking Line [6].

There are two important lessons learnt from the project. First, the project has shown the applicability of Rotor Sail technology onboard a passenger vessel. In this rather special segment of shipping industry, the set of requirements is somewhat different, and stricter than within cargo shipping. On the other hand, from shipowners' perspective the profitability of investment is not solely dependent on the fuel saving potential, but also "secondary" benefits such as brand impacts are also prominent. As such, it is possible that the passenger vessel market takes a role as a fore runner for the wind propulsion market. Indication of this is seen already in Norsepower's latest delivery contract, which has been signed with ferry operator Scandlines.

Secondly, the project clearly shows the potential of Rotor Sail technology on fast vessels. One of the frequent concerns raised is the effect of speed wind, that is considered to diminish the savings potential when vessel is sailing at a speed of similar magnitude as the prevailing natural wind speed. This effect was noted to be minimal. As was seen, the thrust performance on different speed areas was very close to each other, but the higher vessel speed improved the savings potential in terms of propulsion power.

5. LIABILITY

The authors have worked in collaboration with all external parties mentioned in the paper to obtain best possible understanding of the methodologies used to estimate savings potential on Viking Grace. However, to ensure the independency of measurements, authors were not involved in actual analysis. The summaries in this paper are provided by the authors and represent their best understanding of measurements which have been conducted. As such, the external parties mentioned are not liable for statements made in this paper.

6. ACKNOWLEDGEMENTS

Authors would like to thank ABB, Chalmers University, NAPA and Viking Line teams for their collaboration in the project and for their hard work to provide as reliable estimates as possible on the Rotor Sail performance.

7. REFERENCES

1. TALONEN Jaakko, 'Analysis of rotor sail effect for Viking Grace', 14 April 2019

2. WECKMAN Joanna, 'ABB Ability™ Data Analytics for Marine Rotor Sail Performance Evaluation Part 2', 08 April 2019

3. TILLIG Fabian, 'Analysis of the Norsepower rotor sail on board the cruise vessel "Viking Grace" with use of the Chalmers power and journey prediction model ShipCLEAN', 29 May 2019

4. FMI (Finnish meteorological institute), 2019, Retrieved 02 April 2019 from: https://ilmatieteenlaitos.fi/havaintojen-lataus#!/

5. IMO, '2013 Guidance on treatment of innovative energy efficiency technologies for calculation and verification of the attained EEDI', 17 June 2013

6. MORELL Maxime, GUILLAT Solene, 'Survey concerning the installation of Rotor Sail by Norsepower onboard Viking Grace', 26 October 2018.

9. AUTHORS BIOGRAPHY

Ville Paakkari works as a Sales Engineer at Norsepower Ltd. He is responsible for performance prediction of the Rotor Sail technology.

Otto Valkeisenmäki works as a Design and Certification Engineer at Norsepower Ltd. He is responsible for performance analysis and technical calculation.

Wind Propulsion, 15th -16th October 2019, London, UK

DIMENSIONING, DESIGN, MANUFACTURING AND PERFORMANCE ASSESSMENT OF OCEANWINGS WINGSAIL ONBOARD ENERGY OBSERVER

N. Sdez and **M. Van Peteghem** VPLP design, France

SUMMARY

Based on the experience gained through the design and realization of a first prototype, VPLP design engaged itself in the design and making of wind propulsors called Oceanwings®, dedicated to the boat called Energy Observer. The dimensioning, design and making of such a system is documented. The data collected by the embarked calculator are collected, processed and presented. A comparison is made with regards to models used at design stage, together with a commented global energy performance assessment.

NOMENCLATURE

_left1, _right2	Relative to one Oceanwings's position
α	Wing Angular Position (°)
$\beta 0$, $\beta 1$	Camber Angle at bottom and top (°)
AoA	Angle of Attack (°)
AWA	Apparent Wind Angle (°)
AWS	Apparent Wind Speed ($m.s^{-1}$)
BS	Boat Speed (kts)
CoE	Center of Effort
Fx, Fy	Recalculated Forces in ship's FoR (N)
Fx_Th, Fy_Th	Recalculated Theoretical [..] (N)
FoR	Frame of Reference
CFD	Computation Fluid Dynamics
LLT	Lifting Line Theory
P_Tot_Motors	Total Engine Power [kW]
RM	Righting Moment [T.m]
VPP	Velocity Prediction Program

1. CONTEXT

1.1 STAKES

Every second, more than 50.000 ships are sailing on the sea of the world. 90% of global trade transits by sea, it is responsible for a 3-5% of global GHG emissions, and much more when considering fine particules, NOx (18-30%) or SOx (9%) while being bound to double by 2050. While maritime sector is excluded from the Paris Agreement as per its inherently international nature, IMO resolutions 2030/2050 aim at reducing GHG emissions by 50% by 2050 [1], which may let one think that the wake-up call has been heard.
Regulatory measures at regional level are multiplying as well such as the Emissions Controlled Area (ECA) zones, with new possible zones (Mediterranean Sea) or the classification of European S-ECA in N-ECA as well on the 1st of January 2021 are positive signals.
Yet this is far from enough while on the other hand IPCC 1.5 is indicating we have only 12 years until we reach that seriously challenging and potentially catastrophic benchmark and the report states a minimum of 45% CO2 reduction will be required by 2030 and 100% by 2050.
Coercitive as well as incentive measures already exist and in the mean time public awareness is growing. Ship owners increasingly see a true advantage in having a greener image, recent announcements by Maersk which wishes to fully decarbonize its fleet by 2050 or CMA CGM renouncing to use the Arctic North-West route [2],[3].
Let's hope that time is short before ship owners and operators will have to decarbonize their fleet not only to gain market share but to avoid losing ones or being subjected to increased taxes.
A much feared carbon tax on the sector, seemingly unavoidable at short to medium term, is objectively rendered possible by the mandatory Monitoring, Reporting, Verification (MRV) implemented in January 2018. It might be delayed or come with a Slow Steaming international limitation, a position countries like France are advocating for.
That regulatory novelties come with new tools or index such as the EEDI or SEEMP ; that are intrinsically vertuous although they do not cope with all problematic or benefits that come with techniocal solutions, namely with wind propulsion technologies. Severe limitations are indeed coming with the EEDI (e.g. routing and/or variable speed) or with the consideration of the global wind specification MEPC 62-INF.34.
If economic gains directly linked with reduced fuel consumption are obvious, less obvious is the relative uncoupling it allows between transportation costs and fuel costs.
In a general context of ever greater climate emergency, Policy, Price, Perception, Providers and People are all starting to align.
Auxiliary (or even Primarily) Wind Propulsion is part of a set of devices called Energy Saving Devices by the IMO. It is a credible, viable and increasingly profitable option and the industry is taking notice while numerous uncertaincies lay regarding alternative sources of energy onboard ship (ammonia,..) [4]. Wind Propulsion Technologies are as per today the only ones to offer double digits fuel savings.
Among these technologies are the propulsion wings such as the Oceanwings®, this particular technology is promising. Although showing smaller lift coefficients than Flettner Rotors and a lift per unit of surface smaller that Kites, they come with many advantages such as :

- Possibility to multiply systems with very limited harmfull interactions
- Very high aerodynamic lift to drag ratios, allowing for sailing in small apparent wind angles typical of wind assisted ship
- High versatility to wind conditions
- Easy installation and automation

These properties, among others, make it theoretical the most interesting system according to number of studies such as [5],[6] while gains related to other systems either theretical or existed ones are limited or non demonstrated, see [7],[8],[9].

In any case, according to [5], In 2030, the market potential could amount to around 3,700–10,700 installed systems on bulkers and tankers, associated with approximately 3.5–7.5 Mt CO2 savings and 6,500–8,000 direct and 8,500–10,000 indirect jobs.

1.2 OCEANWINGS®

VPLP design got interested vey early to rigid wingsails designs as the design studio participated in the design of BMW Oracle, winner of the 33rd America's Cup back in 2010. That early interest carry on through collaborations with Artemis Racing and Team Hydros for the Little America's Cup in 2013. The studies carried out in that frame allowed for acquiering a true expertise in wingsails design as well as building a strong belief that this type of rigging has a future on other types of boats.

The impossibility to reduced surface area, that is to say, to furl and lower as one can do on a traditionnal rigging consitutes the main reason why it did not propagate.

With the Oceaninwgs concept, VPLP design is aiming at promoting a wind propulsion system that is safe, simple and automated. A new concept of rig, furlable and lowerable, was thus developed on the basis of the high efficiency aerodynamics of multiple élément profiles used on the AC.

Financial aid from ADEME, France's environment and energy management agency, enabled VPLP Design to produce an 8m fully-working prototype, validate feasibility, prove the systems work and acquire sailing data to fine-tune performance models developed in-house. This work leaded to a paper published at the french marime association ATMA in 2017, see [10].

Entirely automated, self-supporting and rotating 360°, Oceanwings® adapts its angle of incidence to the vessel's point of sail to ensure optimal propulsion. Power is managed by trimming camber and twist. Not only does it provide exceptional control, it's also efficient to the point of halving the surface area required to propel a vessel under conventional sail.

Oceanwings® is all about making shipping more profitable and more environmentally responsible. Even though wind propulsion achieves double-digit fuel savings its use cannot give rise to additional and specialist crew. In other words, automation has to be an essential feature. By integrating manufacturing considerations early in the design stage, VPLP has managed to ensure that costs are on a par with the purchase of a high-performance rig and can be quickly recovered by commercial operators. Oceanwings® is not going to be compatible with every kind of vessel or with all the regions of the world. Nevertheless we are convinced that it has a role to play in recreational boating, yachting, shipping (bulk carriers, chemical tankers, oil tankers…) and deep-sea fishing.

The installation of a first industrial demonstrator a mere two years after first sea trials were carried out on the prototype is confirming a Technology Readiness Level (TRL) of 7 in 3 years and largely participates in preparing the higher scale commercial applications.

1.3 ENERGY OBSERVER

After an outstanding racing career under the names of Formule TAG, Tag Heuer, Enza New Zealand, Royal & SunAlliance, Team Legato, or Daedalus, this legendary boat has a new life today under the name of Energy Observer. Built in Canada back in 1983 by naval architect Nigel Irens, under the supervision of Mike Birch as the future skipper, he is by excellence the catamaran that had a decisive impact on the future evolutions of multihulls. Originally being 24.38 meters long, it has been lengthen four times and today is 30.5 meters long and 12.80 meters wide .

A true experimental laboratory with a goal of energy autonomy, Energy Observer is the first vessel in the world, able to produce its hydrogen from seawater, thanks to renewable energies.

Since the very beginning of its new life, VPLP has been involved in the architectural modifications of the boat, that work culminating by the installation of the wind propulsor Oceanwings® first trimester of 2019. The work undertaken by VPLP have had two phases, one beginning of 2016, the other, mid-2018. The first intervention consisted in assisting the yard for the main works. The installation of such many equipments resulted in increasing the displacement so that we had to boost floatability by fitting externally on the hulls foam formers and to evaluate the impact of such modifications on hydrodynamic drag and hence consumption of the boat. Additionnally, to optimize solar panel surface, VPLP designed and dimensioned the cantilever beams supporting the solar panels aft of the wheelhouse.

The second consisted in dimensioning and detail design of specific Oceanwings® to fit the boat's architecture and the energy available onboard. Eventually it was opted for two Oceanwings® with a projected surface of 31.5m² each fitted on one of the float of the catamaran. Controlled by a specific software realized by VPLP and control algorithm based on a pre-calculated database of trim settings, these wind propulsors will be connected to the energy network onboard.

Featuring a reduced energy demand, the two Oceanwings® will on the opposite, depending on wind conditions, allow :

• Increased boat speed, together with the electric motors
• Reduced energy spending, when they reduce loading on the electric motors
• Increased energy production while at sea though hydrogeneration (using electric motors as hydrogenerators)
• Production of hydrogen while at sea using electrolysis thanks to the energy saved or even indirectly produced by their use

If Energy Observer is a true precursor, the use of the Oceanwings® onboard prefigures the way VPLP sees their use : as an auxiliary wind propulsor used in a diversified mix of energy sources onboard ships. Last but not least, it is the project in which we co-designed, fabricated and assembled hand in hands with our industrial partner CNIM.

2. DESIGN AND DIMENSIONING

Design and dimensioning of a wind propulsion system is inherently linked to the platform is will be installed on and are by nature iterative processes, displayed thematically here below, no temporal or sequential logic is underlying.

2.1 ADAPTATION TO SHIP

Adapting a wind porpulsion system to any ship means generally dealing with 3 aspects :

1. Architecture, meaning integration checks with regards for geometrical, operationnal, regulatory and energetic constraints
2. Structure, meaning checking the capacity of existing structures to withstand stresses or accordingly dimensioning reinforcements
3. Performance, meaning evaluating the propulsive contribution of the system per se and with regards to plateforme limitations (structure, stability, leeway, helm balance)

In reverse order,
Building on previous works by VPLP design, a complete CAD model was available. That allowed us to carry out Computationnal Fluid Dynamic (CFD) tests for a range of displacements, speed and leeways. These numerical results were later compared to sea trial tests carried out by Energy Observer's crew, yielding to a hydrodynamic appended hull model. Together with a rudder model – itself based on a potential panel code, a windage model – based on a proprietray wind tunnel database and a wingsail model – which will be later presented, the appended hull model was integrated in a Velocitiy Prediction Program (VPP).

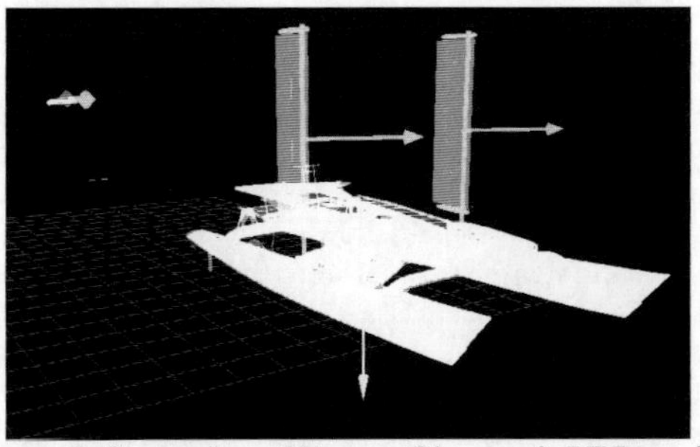

Figure 1. Screenshot of Energy Observer's VPP model

This model allows for evaluating différent configurations against the performance requierements expressed by the crew.
Two configurations were investigated, one featuring two smaller lateralized wingsails and one featuring one central wingsail. That configurations are displayed here below :

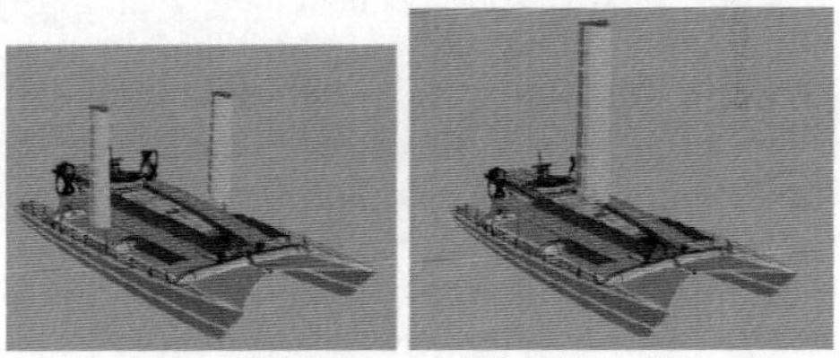

Figure 2. Energy Observer with 2x 31.5 wingsails or 1x 51.5m² wingsail

Structurally speaking, the Oceanwings® is essentially a cantilever beam simply supported in two points. Equipped with auto-aligning bearing, it behaves mostly like a rudder turned upwards. Local stresses are easy to compute and distribute to surrounding structure given that they propagate in-plane. What was sought after was a check regarding global stresses.
The problem is that no structure drawings still exist (if they ever existed), work with GSea Design was thus carried out to model the structure through a unusual method. Once the platform was modeled as per a stick model, it has been assigned inertias based on geometry and estimated stiffnesses, that have been compared with a database of same geenration catamarans : the underlying hypotheis being that if the stiffness distribution is correct, then stress distribution will be as well.
Traditionnally, offshore race catamarans are dimensionned under a load case called GH1 meaning the leeward hull plunging. That load case was applied to the stick model in accordance with a 18T displacement and a 60T mast compression (corresponding a RM of 115 T.m and 450m² of conventionnal sails up). The resulting load distribution was compared to that regulatory load cases of a Motor Yacht (DNV GL – HSLD Pt3 Ch1 et Bureau Veritas - NR 500 DT R01 E.) applied on the 33T current catamaran. This study showed a risk under the BV Broadching load case, the load case has limited probability as long as boat speed are limited.

To the stick model was later added the local Oceanwings® ultimate loadings, which showed little difference in bending and torsionnal moments on the middle arm. The stresses do not overpass 6.2T at upper bearing.
The wingsails shall remain the weak point in the chain of stresses, local reinforcements were thus calculated on the basis of the ultime load cases.
What appeared on the central wingsail confiuration is that the nacelle's structure was unable to cope with the loadings and requieres rigging lines to sustain the upper bearing.

Architectural and visual considerations were of major concern. A central installation would have meant, besides the structural arrangement mentionned above, a drastic modification of the yacht's interiors, which are entierely equipped with linings. A lateralized installation instead, besides being structually simpler had the advantage of putting the rig's systems in the technical zones.
That configuration meant reduction or splitting of capacities that were in the way. It induced greater complexity on the wings structure itself though as it requiered a longer portion of « naked » tube in-between upper bearing and aerodynamic surface. That is because the wingsail has to turn 360° above the horizontal solar panels visible in green in figure 3.
For visibiliy reason, the light disposition had to be changed and head lights were put on the masts.
Finally, all actuators have been designed to be powered by 400 VDC boat's energy bus while drives and computers were put on a stabilized 24 VDC network itself powered on the 24 VDC boat's system bus.

2.2 SPECIFIC ADAPATATIONS

The aerodynamic polars of a wingsail of finite length made of multiple element profiles are calculated with a non-linear Liftine Line (LLT) model, including wind gradient and ground effect. That design tool, developped in collaboration with Kiel University has been tested versus CFD and wind tunnel tests [11], see also [12] ,[13]. Alimented with 2D CFD results from calculations ranging upon AoA and camber angles, that rapid design tool (about 2 seconds per calculation point) allows from complete aerodynamic polar generation.
Coupled with an optimizer with a constraint on bending moment at upper bearing, it allows for calculation of optimal settings and propulsion with regard to internal structure limitations.
Once general dimensions are validated against the ship's performances, a profile optimisation is carried out to select for best profiles and relative profile positions (generally called gap and overlap) for a given Reynolds number, itself corresponding to average AWS at VPP design points.
One point is of special attention, as the wingsails span is of 10m and the catamran's airdraft is of 14.5m, the wingsails work in the Earth's wind gradient. This is aerodynamically an issue as, when computing the apparent wind speeds and angles along span, the latter greatly varies ; especially for slow boats and following winds.
In order to adapt locally to a variable effective AoA, one has to realize an aerodynamic twist, this is realized by setting differently the camber angles at bottom and top, see Figure [9].
In our case, that is originally done through structural twisting of the secondary mast. Equipped with a lug, it drives the ribs as it rotates. Designed in carbon fibers, it is very stiff regarding bending granting near-constant gap and overlap values but relatively flexible torsionnally. Not enough to deflect under aerodynamic loadings but not to much to generate high forces on actuators. This is realised by carefully selecting fiber type and orientation.

3. REALIZATION AND TESTING

3.1 SYSTEM DESCRIPTION

Each Oceanwings® is a system comprising :

- A carbon fiber support structure :
  - 1 main and 1 secondary masts
  - 1 lower and 1 upper boomboxes
  - 7 Main Element Ribs
  - 12 Second Element Ribs
- A reefable sail plan of a 31.5m² projected surface
- A set of electromechanic actuators :
  - 1 geared motor on a slewing ring
  - 2 geared motors on pulleys
  - 2 motors on cylinders
- A set of control system / support devices :
  - 2 electric cabinets per hull
  - 2 rotating electric cabinets
  - 1 slipping ring
  - Sensors(Wind vanes, Strain gauges ,..)
- A ship interface :
  - 1 upper and lower autoalignig bearings
  - Electric plugs

Mutualized equipements :

- GPS / Compass
- Fixed windvane
- PC + I/O modules
- Control Touch screen

3.2 TESTS, INSTRUMENTATION, INSTALLATION

Subsets were tested at factory (FAT) and later assembled and tested at CNIM facilities in la Seyne-sur-Mer. That process allowed for a. unitary mechanic and mechatronic tests, b. full system testing at workshop, c. control/command testing, d. instrumentation of masts.
The main masts were indeed fitted with strain gauges on 3 axes (longitudinal and lateral bending and torsional moment). Strain gauges were calibrated in workshop.

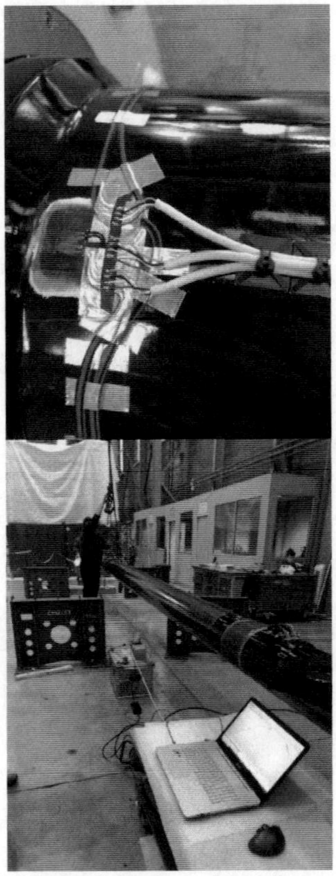

Figure 3. View of gauges and bending test set-up

Masts are put under pure stresses on each of the above-mentionned axis ; a computer simultaneously records gauges signals and applied force through the use of a electronic dynamometer. When well executed, very nice slopes are obtained, allowing for slope coefficient extraction. Thus, an image of the strain levels in the masts is correlated with bending moments. Once the masts installed, the gauges are set to 0 when the boat is docked and subjected to no wind.

Figure 4. System installation

The installation itself was performed on the 18 of April 2019 in Amsterdam. It consisted simply in a crane operation, a fixation of bearing and a pluging to équipements per-installed onboard.

The Oceanwings® were packed on a truck at 9am, Energy Observer was sailing at 7pm in front of Amsterdam railway station. The presence of the fenders indicates the suddenty of the trials that were planned on the next day.

Figure 5. First Trials at "sea"

4. PERFORMANCES

4.1 FOREWORD

Having instrumented the Oceanwings® allow for validating the aerodynamic models used at design stage against reality. It implies the the handling of time series generated by the gauges on one side, the ones generated by windvanes, GPS and compas on the other ; changing of reference frames (Fluid FoR – Wing Aero FoR – Gauges FoR – Ship FoR) ; recalculating stresses using design stage models for each instantaneous environmental configuration.

Logs are recorded every second and are comprised of around 90 variables. The here-under presentation of results present the sole one way of Energy Observer from Tromso to Spitzberg. The reason that leg has been chosen is because with other 500 nm it corresponds the longest voyage in distance in duration in open seas. The database use is comprised of about 604.800 logs corresponding to 7 days from the 2nd to the 9th of August 2019.

4.2 AERODYNAMIC MODEL CHECK

Before explaining our process, it must be said that the catamaran is not equipped with gyrometers/accelerometers, that proved to be quite an issue in the data assessment as inertial phenomena, namely due to swell. Gauges signals are very noised. As swell periods are less than 15 seconds in the north of Norway [14], a minimal averaging time of 4 periods has been set, hence 60 seconds averaging time. It allows for good capturing of tendencies on lateral gauges (2) but greatly decreases the values recorded on longitudinal gauge (1). That fact has an impact on presented results.

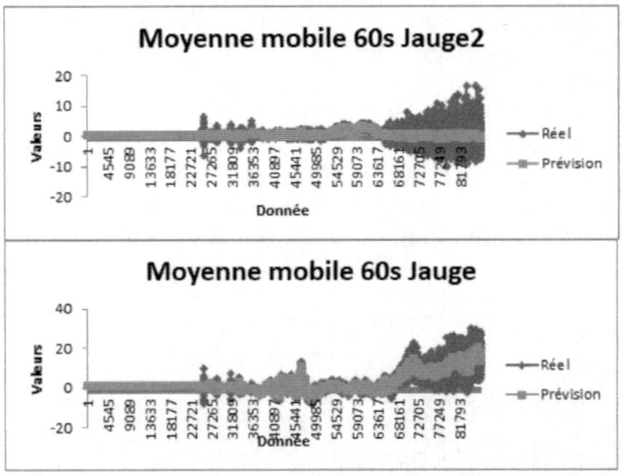

Figure 6. Raw and averaged time series of longitudinal (2) and lateral (1) gauges

Another issue that was faced during the processing of logs was that portside wingsails's gauges were damaged – most probably due to water infiltration. A first one occurring the 15th of July (dark blue corresponding torque gauge on the left figure), a second one occurring on the 7th of August (light blue corresponding lateral and longitudinal gauges on the right figure), a zero deviation is clearly visible, which is typical in that case.

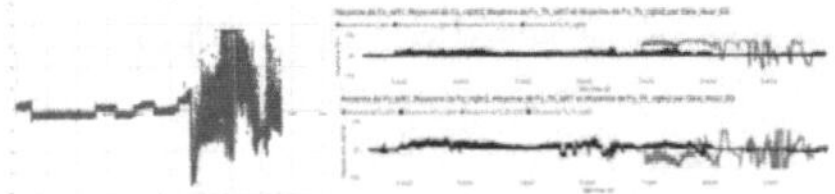

Figure 7. Zero deviation of gauges

The measures of the gauges (mV/V) are converted to moment (kN.m) and efforts, under the hypothesis of an aerodynamic CoE and then projected on the ship's FoR, yielding to Fx, Fy.

On the other side, the LLT results must be recomputed for every single set of conditions for comparison. Despite the fast computation of the LLT such computations would have proved prohibitively long.

Taken from the matrix of optimized settings, a simple response surface was extracted in the form of :

$$Cl,Cd = f(AWA) = \sum_{i=0}^{i=6} \lambda_i AWA^i \quad (1)$$

Projected on ship's FoR to obtain Cx, Cy and then multiplied by 60sec averaged measured AWS, they yield to Fx_Th, Fy_Th – theoretical values of efforts according to our models.

That approach yield very good results in terms of Fy and fair ones in terms of Fx, given the above-mentioned issues on averaging gauges, correlation is presented below:

Correlation	*Fx_right2*	*Fy_right2*	*Fx_Th_right2*	*Fy_Th_right2*
Fx_right2	100%			
Fy_right2	46%	100%		
Fx_Th_right2	70%	69%	100%	
Fy_Th_right2	44%	97%	61%	100%

A statistical visual representation is added to help one visualize:

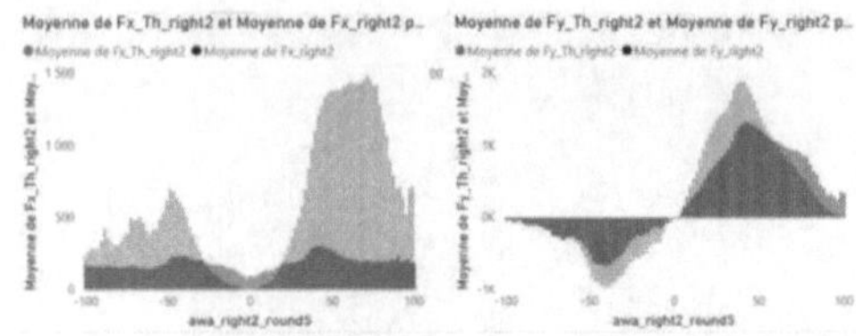

Figure 8. Comparison between Fx, Fy, Fx_Th, Fy_Th

Results along the x axis are less accurate although well correlated meaning tendencies are captured although values are not accurate. Heavy suspicion lies on the fact that the averaging in longitudinal gauge leaded to that situation.

4.3 CONTROL LAWS CHECK

The Oceaniwngs system is fully automated, in practical a database of optimal trimmings generated by the VPP is interrogated with regards to any current environmental conditions and a command is generated accordingly to move the actuators. These allow for 3 movements, on general rotation (α angle) corresponding with AoA, two cambering commands (β angles) generating camber and twist.

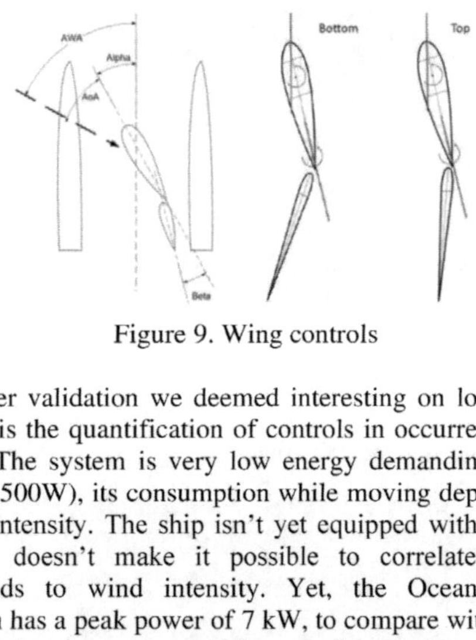

Figure 9. Wing controls

Another validation we deemed interesting on long time series is the quantification of controls in occurrence and time. The system is very low energy demanding while still (<500W), its consumption while moving depends on wind intensity. The ship isn't yet equipped with shunts, which doesn't make it possible to correlate power demands to wind intensity. Yet, the Oceanwings® system has a peak power of 7 kW, to compare with 2x 41 kW of electric motors used for propulsion onboard.

The charts here-under represent they time evolution of variables (only one hour represented for visibility). From left to right are the measured instantaneous AWA, the computed AWA_mean (corresponding to the integration of local AWS with regard to a wind gradient power law) and α angles closely related to AoA. Averaging and filtering are visible.

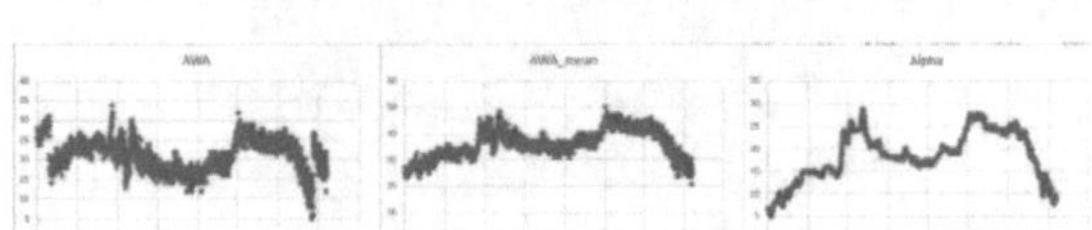

Figure 10. Time series of AWA, AWA_mean, Alpha

The number of occurrences of a moving sequence, its duration and the duration between occurrences are also important values to have a look at, on the 7 days voyage.
Port and starboard side respectively, 3514 and 2288 activations were recorded, with mean duration of still phases of 260 and 424 seconds, and mean duration of moving phases of 1.6 and 1.7 seconds.
The differences between regulations of left and right system is due to the fact the left wing was mainly on the leeward side, hence evolving in more disturbed flow.
Obviously, these values correspond to a low energy control strategy, more frequent controls are of course possible, namely in manoeuvring.
Finally, to be noticed is the fact that during the 7 day voyage considered, manual mode was never used, meaning the computer has had full control of the system.

4.4 PERFORMANCE ASSESSMENT

That part was the last one to be explored because it involves the highest number of sub-models and assumptions. What appeared and is very interesting is that not only is the boat's behaviour complex because it has an hybrid aeromechanical propulsion, but also because it has its own global energy management system. The analysis must then take into account not only the propulsive contribution of the engines, that was expected, but also the control strategy implemented at the time of measures, that was unexpected.
Data analysis is still being carried out and the short time allowed to write that paper did not allow to go back to previously presented results in the light of that finding.
Let's discuss that finding looking at the following time series :

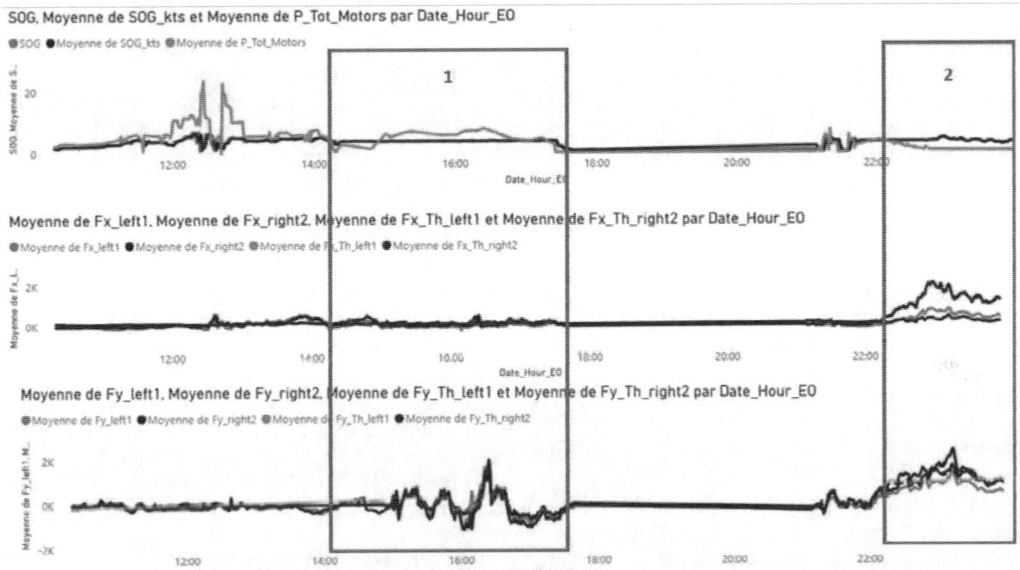

Figure 11. Time Series (P_Tot_Motors, BS, Fx, Fy, Fx_Th, Fy_Th

Zone 1 corresponds with a speed regulation strategy. Motor power (orange on top chart) is regulated so that speed (dark blue on top graph) is constant. Fx curves (central graph) and engine power are inverted (left peak on central graph corresponds with a off-peak in engine power).
Zone 2 corresponds with a power regulation strategy. More precisely, we can see engine power decrease before being set constant. That strategy is typically adopted at night as it allows for prediction of energy levels at sun rise next morning. Motor power is set to a mere 700W, just enough for the propellers to compensate for their own drag. One can see that speed varies visibly (top dark blue curve) and accordingly with Fx curves (central graph).
One can also see that differences between measured Fx, Fy and recomputed values Fx_Th, Fy_Th are visibly dependant and these zones.

Actually, the current state of our understanding is that it is not inherently the control strategy that is of question but the share of aeolian propulsion over mechanical. As the former gets larger, the boat behaves more like a sailing yacht, being subjected to time dependencies at sea (wind, waves) hence making predictions more difficult.
Data are still being processes by Energy Observer as well but what is now certain is that while having sailed more than 3000 nautical miles from Saint-Petersburg to Spitzberg, the Oceanwings® have assumed more than 42% of propulsive power.
That figures rises to 83% on the specific voyage Tromso-Spitzberg discussed in this paper (837 kWh as per engine equivalent energy), which was subjected to a very low cloud cover, making solar power less available.
But these figures, said quite brutally, are meaningless by themselves. They are dependant on boat speed, share of aeolian over mechanical propulsion and global

propulsive efficiency. What is of interest on the opposite is knowing that the predictions of thrust and lateral forces are well modelled – at least in certain configurations. And that these configurations will be those of interest for higher scale applications where aeolian propulsive power to engine power will probably often be superior with respect to the one on Energy Observer, as per a Wind Assisted Ship Propulsion (WASP). Having been able to qualify our models against reality allows for a always higher level of confidence in future numerical simulations during design phases.

5. CONCLUSIONS

A design and dimensioning of aeolian propulsion system, answer to a wind propulsion assistance need onboard Energy Observer has been done under numerous and various technical and economical constraints. The delivered system Oceanwings® 3.2 corresponds with the industrial demonstrator phase and includes specific developments or features. Limited integration works and easy installation were key elements of success for this project.

If still a lot remains to be learnt on the behaviour of hybrid and globally controlled ships, the current state of log analysis on a large dataset are satisfying and hence constitutes a good basis for model updates.

Fixing of gauges, further ship instrumentation and interfacing with the Oceanwings® will allow for further analysis under different scenarios.

To come in 2020, trans-Atlantic and trans-Pacific voyages will be a very good opportunity to generate data and help understand the physical reality of the system, the ship and their functioning.

6. REFERENCES

[1] Third IMO GHG Study 2014; International Maritime Organization (IMO) London, UK, April 2015; Smith, T. W. P.; Jalkanen, J. P.; Anderson, B. A.; Corbett, J. J.; Faber, J.; Hanayama, S.; O'Keeffe, E.; Parker, S.; Johansson, L.; Aldous, L.; Raucci, C.; Traut, M.; Ettinger, S.; Nelissen, D.; Lee, D. S.; Ng, S.; Agrawal, A.; Winebrake, J. J.; Hoen, M.; Chesworth, S.; Pandey, A.

[2] Website : https://www.maersk.com/news/articles/2019/06/26/towards-a-zero-carbon-future

[3] Website : https://www.tradewindsnews.com/offshore/imo-france-proposes-mandatory-slow-steaming/2-1-585206

[4] Jean-Pierre Husson, Clément Ramos, Stéphane Amant, Carbone4-Bertin (2019), Le secteur maritime navigue-t-il vers la décarbonsation ?

[5] CE Delft - Study on the analysis of market potentials and market barriers for wind propulsion technologies for ships

[6] Marten Silvanius, (2009) Master Thesis, Wind assisted propulsion for pure car and truck carriers

[7] van der Kolk, Nico & Bordogna, Giovanni & Mason, James & Despraires, P & Vrijdag, Arthur. (2019). Case Study: Wind-Assisted Ship Propulsion: Performance Prediction, Routing, and Economic Modelling.

[8] 4th Conference on Ship Efficiency Hamburg, 23-24. September 2013 Enercon E-Ship 1 A Wind-Hybrid Commercial Cargo Ship

[9] J. Väinämö, 24th International HISWA Symposium on yacht and naval construction, Operational expériences and results from the first reference installation from Nov-2014, Roro-ship Estraden (9700 dwt)

[10] Nicolas Sdez, Marc Van Peteghem, Aile de propulsion : du grémeent performance à la propulsion auxiliaire de navire, ATMA (2017)

[11] K. Graf, A. v. Hoeve, Simon Watin The third International Conference on Innovation in High Performance Sailing Yachts – Comparison of Full 3d-Rans Simulations with 2d-Rans / Lifting Line Method calculations for the flow analysis of rigid wings for high performance multihulls

[12] Viola, Biancolli, Sacher, Cella, A CFD-based wingsail optimisation method coupled to a VPP, 5th HPYD, Mars 2015

[13] Chapin, Gourdain, Verdin, Fiumara, Senter, Aerodynamic study of a two-elements wingsail for high performance multihull yachts, 5th HPYD, Mars 2015

[14] Reistad, M., Breivik, Ø., Haakenstad, H., Aarnes, O. J., Furevik, B. R., and Bidlot, J.-R. (2011), A high-resolution hindcast of wind and waves for the North Sea, the Norwegian Sea, and the Barents Sea,

[15] MEPC.245(66) - 2014 Guidelines on the method of calculation of the attained Energy Efficiency Design Index (EEDI) for new ships, as amended

[16] MEPC_282(70)_2016 Guidelines for the development of a SEEMP (including Corrigendum

7. ACKNOWLEDGEMENTS

Special acknowledgments to the Energy Observer's crew for their help in collecting relevant data.

Acknowledgements to all of our partners and providers who helped us deliver the system.

8. AUTHORS BIOGRAPHY

Marc Van Peteghem holds the current position of president-founder at VPLP design. VPLP became famous first in multihull design their boats won many races and records including round the world Jules Verne trophy, every Route du Rhum since 1990, in 2010 the 33d Americas Cup ORACLE.

Nicolas Sdez holds the current position of engineer at VPLP design. He is responsible for the development of the Oceanwings product since 2016 as technical expert and project manager. His education includes naval architecture and aeronautics.

SHIP DATA DRIVEN PROPULSION MODELS & WIND-BASED PROPULSION TECHNOLOGY

J. E. Buckingham, D. R. Pearson and **E. Storey,** BMT Defence and Security UK Ltd, UK

SUMMARY

This paper presents the approach taken to develop and align a parametric ship model to metocean data and known ship performance data measured at sea, and the use of this aligned baseline model in conjunction with first principle models of wind-based Energy Saving Technologies (EST) to derive the benefit of such technologies over specific voyages. The EST compared in this paper are:

- Flettner rotors
- Wingsails
- Turbosails

A specific voyage conditions are studied for a given ship in order to provide a broad and representative range of seagoing conditions for the vessel, with the model then being used to derive the relative fuel savings from the EST. This output can then be run-through a techno-economic analysis to derive the monetary savings and inform a potential business case for installation of such EST. The vessel presented in this paper is a 61,000 dwt bulker, which has served as a demonstrator vessel in previous BMT publications and is a well understood baseline for comparison.

1. INTRODUCTION

The availability of Automatic Identification System (AIS) and other on-board ship data has opened up new possibilities with respect to the development and application of active ship propulsion models. Such models mimic the parametric behaviour of the propulsion system at regular time intervals. Both traditional and modern parameter-prediction techniques are used to confirm alignment between the computer model and the ship data for a range of sea voyages in different sea states.

Wind-based technologies may appear to be a panacea for free propulsion, but the world's oceans do not always experience wind of appropriate speed at the right heading relative to the ship. Key operating areas and routes are identified for the successful operation wind-based EST.

The ETI-funded, VTAS project has combined large publically-available metocean datasets with on-board performance data to build the best possible model definition for a specific vessel and the seagoing conditions experienced on a specific voyage [1]. Vessel modelling has achieved alignment for several ships across a range of ship types.

An aligned ship model can then be used to undertake retrospective analyses of actual ship operations with the addition of single EST or combinations thereof such as Flettner rotors, Wingsails and Turbosails. Examples of these are demonstrated for a 61,000 dwt ship. These technical studies provide the foundation necessary to support a wider techno-economic assessment which allow the feasibility and economic viability of an EST installation to be understood.

2. DATA HANDLING

2.1 SHIP DATA AVAILABLE

Ship data refers to data recorded on board a vessel by some form of Fleet Performance Management (FPM) System. Depending on the system this can encompass a wide range of navigation and vessel performance variables taken at intervals as frequently as 1 minute, including:

Navigational	Machinery	Meteorological
Vessel location	Shaft revolutions per minute (rpm)	Wind Speed & Direction
Time & Date	Shaft power	Significant Wave Height (SWH)
Course over Ground (COG)	Main Engine fuel mass flow	Current Speed & Direction
Heading	Non-propulsive fuel mass flow	Air temperature
Speed Through Water (STW)	Draught fwd & aft	Water temperature
Speed over Ground (SOG)		Relative humidity

FPM based Ship data is the ideal dataset to use as it generally provides all necessary variables which can be used as a direct input to the vessel model Vessel Energy Device Assessment Tool (VESDA), or can be in turn be used to baseline a ship model when using alternative data sources.

2.2 SHIP DATA UNAVAILABLE

Where ship data is not available to be used directly within VESDA, or the choice has been made to compare performance in times when data has not been recorded, then inputs can be derived from alternative source.

AIS data is available from large public datasets that can provide the following data at time intervals ranging from 5 minutes to several hours:

- Timestamp
- Latitude + Longitude
- Heading
- Speed over ground
- Draft
- Rate of turn

In turn this data can be combined with open source metocean data which provides weather data that is essential to generating accurate ship resistance models:

- Timestamp
- Latitude + Longitude
- Significant Wave Height
- Wind speed + direction

How these datasets combine to provide a modelled result is illustrated conceptually in Figure 1. The details of how VESDA uses this data to calculate fuel consumption is captured in previous publications on the subject [1, 2].

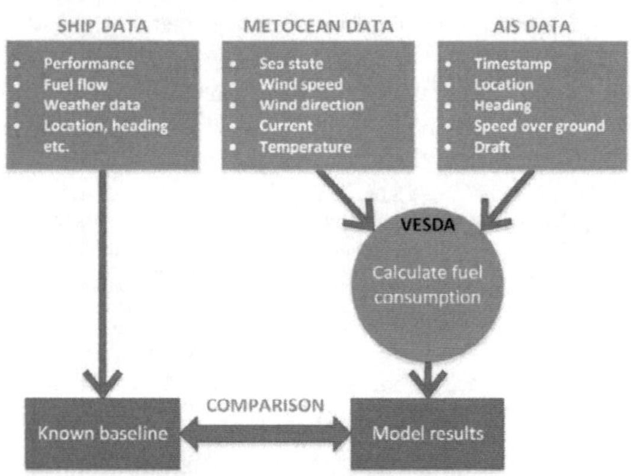

Figure 1: Model data usage

3. MODEL ALIGNMENT

The alignment of the model to the known baseline as illustrated in Figure 1 is conducted by direct comparison of specific key variables in the Ship data and the Model results. One of the key indicators is the shaft power requirements, as illustrated in Figure 2.

A days sailing worth of data has been shown in Figure 2, and the Ship data baseline (red) can be seen to be fairly stable. The model estimate (green) is slightly variable but overall achieves an average deviation of 0.043%. The model estimate with a wind-based EST attached is illustrated in blue.

This type of output can be used to refine the initial estimates used to create the model, for example in this case the hull roughness used to calculated the frictional resistance is likely to be reasonable, however the model may be over-reacting to changes in resistance due to wind and waves, as seen in the peaks and troughs on the green line.

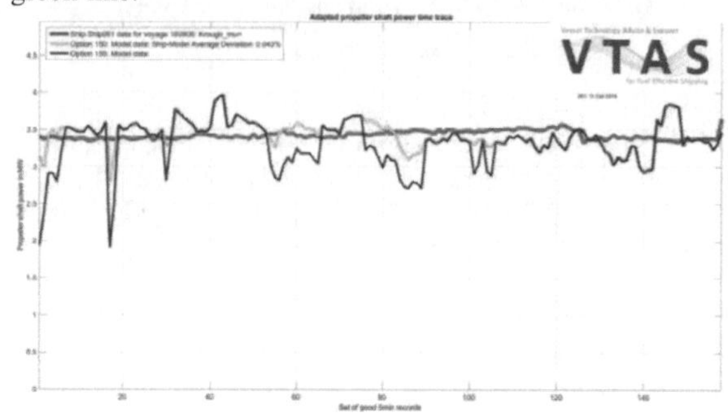

Figure 2: Shaft power time trace

Several different outputs are analysed over multiple different days of sailing for the vessel, in order to capture a representative range of conditions. Once the model achieves and acceptable level of agreement with the Ship data, then it can be used for comparative studies with the addition of ESTs.

4. ENERGY SAVING TECHNOLOGIES

The intention in this scenario is to model the same vessel in the same scenario 3 times, each time with a different type of wind-based EST on board in an appropriate configuration. A basic description of their operating principles and the specifics of the ship fit are outlined below.

4.1 FLETTNER ROTORS

Flettner rotors (FRs) are not a recent invention, having first gone to sea in the 1920's when their potential was realised [3]. FRs are a form of wind-based propulsion that utilizes the 'Magnus effect', a phenomenon exhibited by a spinning body in a fluid flow incident upon it.

A FR typically comprises a cylinder with an endplate affixed to the top, mounted vertically to the deck of a ship. Through the action of a motor, the cylinder rotates in an air stream and a lift force is generated that can contribute to the propulsive needs of the ship [4].

In this scenario, a ship set of two 24m high rotors was chosen, with an aspect ratio of 6:1. The lift coefficient (CL) and drag coefficients are taken from Ackeret's work with the rotors running at 300rpm as described in [4].

4.2 WINGSAILS

Wingsails (WS) operate under the same aerodynamic lift principles as an aircraft's wing, just as traditional fabric sails do. Unlike traditional sails however, due to their comparative rigidity they hold a more efficient (in terms of lift/drag ratio) aerofoil shape.

A Wingsail is typically formed of symmetric aerofoil sections, with some facility built in allowing them to adjust their camber to improve efficiency depending on wind direction.

The WS modelled in this study are based on the "Walker wingsail" 3 bladed type [5], and for this study a ship set of four 19.5 metre sails with an aspect ratio of 3.5 was chosen. The lift coefficient (Cl) given by Walker [5] is 3.0 with a drag coefficient (Cd) of 0.3.

4.3 TURBOSAILS

Turbosails (TS) are described under a variety of names, however for the purposes of this paper they are referred to as "Turbosails" as per the original design by the creators [6].

The operating principle is based on a traditional aerofoil with a very high relative thickness (up to 66%), where to prevent detachment of the boundary layer, a suction is introduced at the aft end of the upper boundary. This suction is created by a fan mounted inside the body of the TS. The suction dramatically improves the Cl and reduces the Cd. This is illustrated in Figure 3.

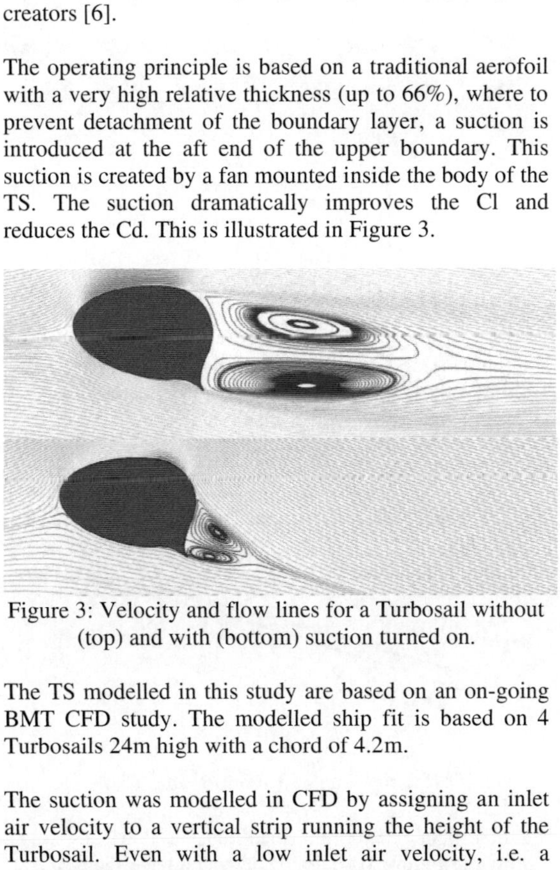

Figure 3: Velocity and flow lines for a Turbosail without (top) and with (bottom) suction turned on.

The TS modelled in this study are based on an on-going BMT CFD study. The modelled ship fit is based on 4 Turbosails 24m high with a chord of 4.2m.

The suction was modelled in CFD by assigning an inlet air velocity to a vertical strip running the height of the Turbosail. Even with a low inlet air velocity, i.e. a suction velocity equal to the surrounding wind speed, the ratio of Cl/Cd increased by 88% between models shown in Figure 3.

The Turbosail angle of attack (AoA) employed here is 6° (Cl=3.793; Cd=0.44) based on the current preferential ratio of lift to drag coefficients. Work continues with a range of suction rates and other AoA with different wind speeds to establish the best AoA setting for a given set of ambient wind and ship operating conditions.

5. MODELLING

As all EST in this study are wind-based, it is possible to represent them in the VESDA model in a similar way, by varying the coefficients of lift and drag relative to the specific operating parameters of the EST themselves, e.g.; rotational speed of the Flettner rotor, AoA for the Wingsail, and AoA and suction coefficient for the Turbosail. Power demands for the FR and the TS fan demand are also included.

5.1 SHIP MODEL

The VESDA ship model has been extensively refined and the methods are discussed in previous publications, and once baselined to a specific ship (see Figure 1) it can be used here to study the effects of external changes such as EST additions, wind, waves, changes in trim, hull fouling, etc. The ultimate goal is to calculate fuel consumption, which broadly operates on the following logic:

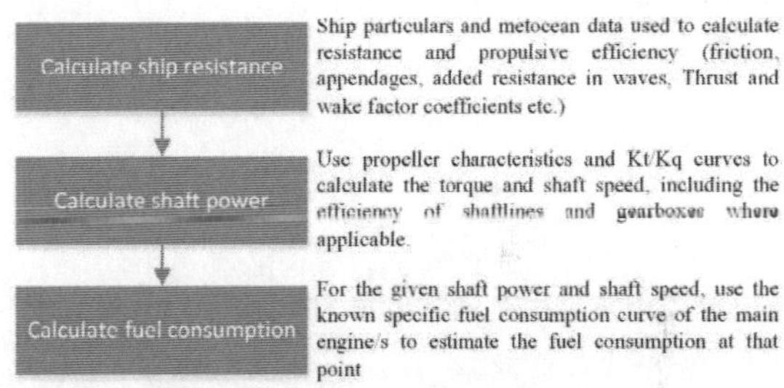

Figure 4: VESDA propulsion model

5.2 EST MODELS

As a vessel is modelled through the voyage, the wind direction and strength is supplied to the EST model at the timesteps available from the metocean data. The smaller the timestep, the higher the "resolution" and the confidence in the results. The EST model will then respond in accordance to that wind input (e.g. change the rotational speed of the FR) and the forward and lateral thrusts are calculated. The forces from wind devices are illustrated in Figure 5.

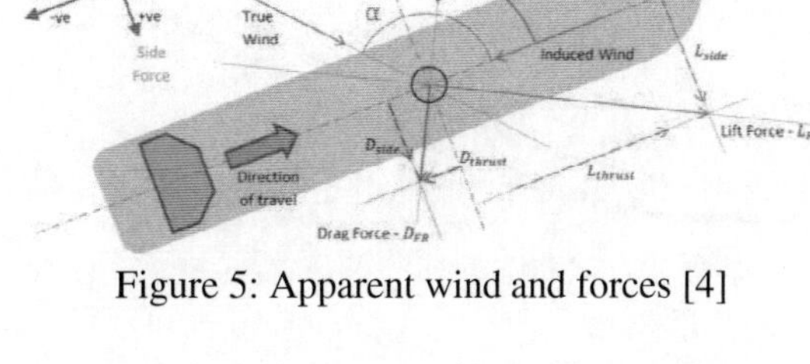

Figure 5: Apparent wind and forces [4]

The forward thrust (L_{thrust}) from the EST is passed back to the Ship model (Figure 4) and manifests as a reduced ship resistance, which in turn reduces the required shaft power/speed. Thus the main engine fuel consumption is reduced.

The lateral forces and drag are used to calculate other factors that affect ship handling and fuel consumption such as increased rudder angle and drag, power required to operate the EST is added to the Ships Electrical Load (SEL), and increased drift and leeway affecting the hull resistance.

4. RESULTS

The following results are for a direct comparison of the EST fits averaged over all wind directions of a sea state 4 for a varying ship speed. These results give an impression of the relative performance of the different EST under the same wind and ship operating conditions.

EST	No. fitted	Sizes
Flettner rotor	2	24m x 4m
Wingsails	4	19.5m x 5.6m
Turbosails	4	25m x 4.2m (chord)

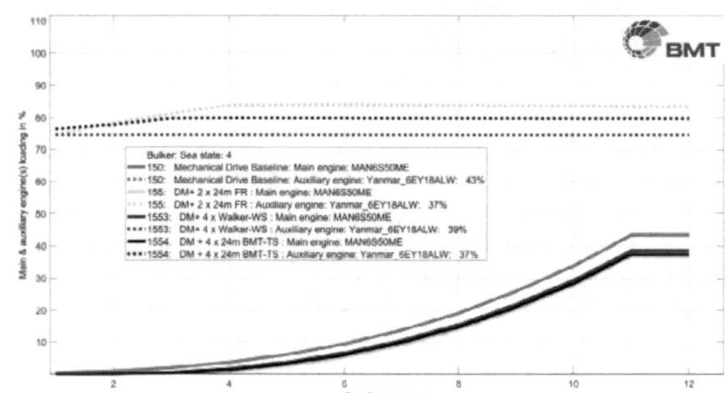

Figure 6: Main and Auxiliary Engine loading ship speed

In Figure 6 the baseline (no EST fitted) main and auxiliary engine loadings are shown in red. The auxiliary loading is coincident with the dotted blue line for Wingsail loading, as the Wingsail does not increase the auxiliary load in weathervane mode.

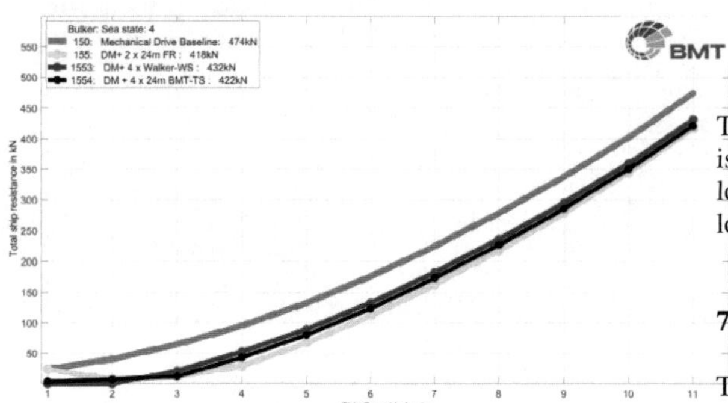

Figure 7: Total ship resistance v ship speed

Figure 7 shows the effective resistance as seen by the propeller. The three EST solutions make a comparable impact on the resistance reduction at sea state 4.

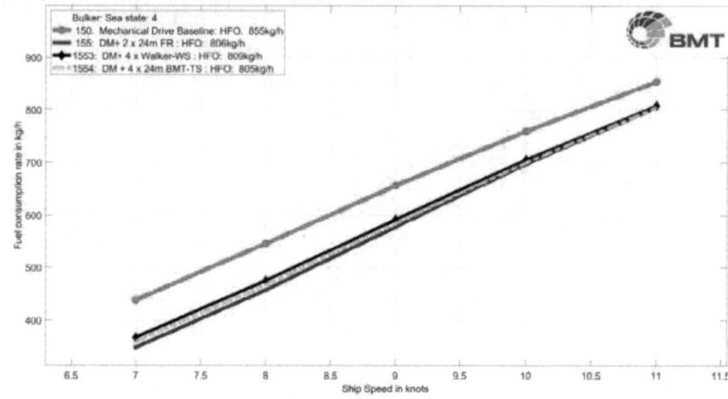

Figure 8: Fuel consumption rate in kg/h

The main engine fuel consumption is likewise reduced as shown in Figure 8.

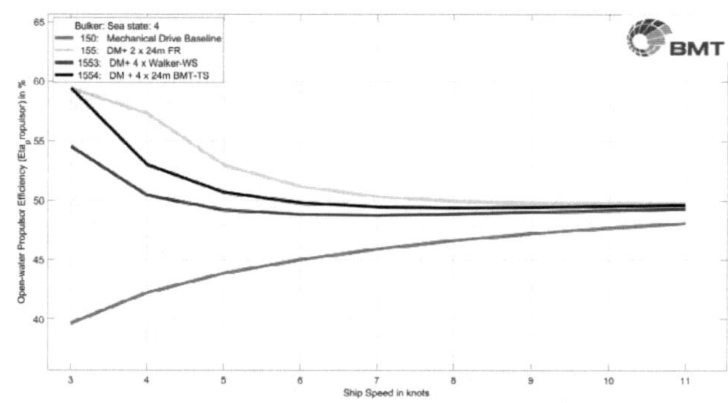

Figure 9: Open water propeller efficiency v ship speed

Figure 9 shows how a lower loaded propeller leads to a better indicated propeller efficiency.

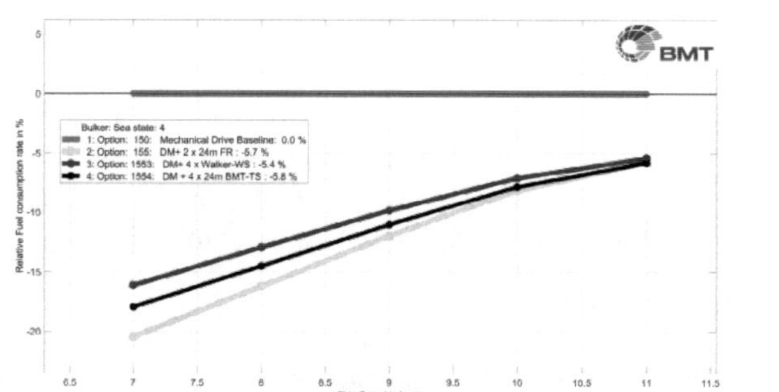

Figure 10: Relative fuel consumption rate v ship speed

The total fuel consumption (main plus auxiliary engines) is shown in Figure 10. The FR offers better savings at lower ship speeds but then the propulsion power is also low in such conditions.

7. CONCLUSIONS

The three design solutions offered here provide a comparable benefit at sea state 4 across the ships operating speed range. At other sea states the relative differences will continue to be small. The main impact on the ship is the deck area consumed by the EST solutions and how that impacts on the requirements for visibility from the conning position on the bridge.

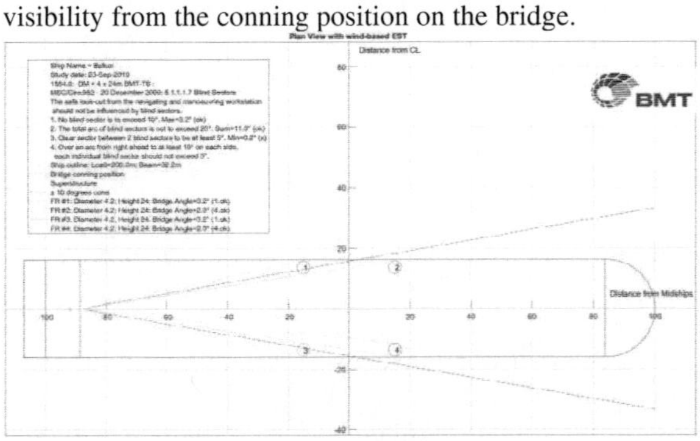

Figure 11: 4 x TS plan view

Figure 11 shows how the four TS in this design almost meet the requirements of the IMO Misc.Circ 982, [7]. The challenge in this case is the necessary angular separation between adjacent TS, a challenge met by the two in number, FR design.

The maturity of the FR design is indicated by a choice of over two suppliers whereas the Walker-WS is not currently commercially available. The TS concept is available through the Econowind Ventifoil design [8] but it is the BMT study design version which has been presented here.

Each of the three wind-based EST designs shown here offer comparable fuel savings. Each ship owner will wish to consider the economic merits of the commercial opportunity offered by EST suppliers in addition to the ship-fit feasibility of the equipment. BMT with Black & Veatch under the VTAS agreement, seek to support ship owners and others when they are considering how best to introduce wind-based and other ESTs to their ships.

8. ACKNOWLEDGEMENTS

This paper represents work carried out by BMT Defence And Security UK Ltd as part of the VTAS joint venture with Black & Veatch, whose support is gratefully acknowledged. The authors wish to thank Will Hopes, BMT for his support.

It is BMT's intention to claim copyright for this work. The kind permission and resources granted to the authors by BMT are acknowledged with thanks. All findings, ideas, opinions and errors herein are those of the authors and are not necessarily those of BMT Defence and Security UK Ltd.

9. REFERENCES

1. J. E. BUCKINGHAM and D.R. PEARSON, 'Modelling The Performance of Energy Saving Technologies on Ships', RINA, Full Scale Ship Performance, 24-25 October, London, UK

2. J. E. BUCKINGHAM and D.R. PEARSON, 'Modelling Alternative Propulsion Technologies for Merchant Vessels', RINA, Power & Propulsion Alternatives for Ships, 23rd January 2019, London, UK

3. L. Bergeson and C.K. Greenwald, 'Sail assist developments 1979-1985'. Journal of Wind Engineering and Industrial Aerodynamics, 19 (1985) 45-114. Elsevier Science Publishers, B.V., Amsterdam. 1985.

4. D.R. PEARSON, 'The Use of Flettner Rotors in Efficient Ship Design', RINA, Influence of EEDI on Ship Design, 24-25 September, 2014. London, UK

5. J.G. WALKER, 'A High Performance Automatic Wingsail Auxiliary Propulsion System for Commercial Ships', Journal of Wind Engineering and Industrial Aerodynamics, 20 (1985) 83-96. Elsevier Science Publishers, B.V., Amsterdam. 1985.

6. B. CHARRIER et al, 'Fondation Cousteau and Windship Propulsion, 1980-1985'. Journal of Wind Engineering and Industrial Aerodynamics, 20 (1985) 39-60. Elsevier Science Publishers, B.V., Amsterdam. 1985.

7. IMO MSC/Circ.982. 20 December 2000. "GUIDELINES ON ERGONOMIC CRITERIA FOR BRIDGE EQUIPMENT AND LAYOUT"

8. Econowind Ventifoil design.

9. VTAS, Vessel Technology Advice & Support.

9. AUTHORS BIOGRAPHY

John Buckingham CEng FIMechE, is the Chief Mechanical Engineer of BMT Defence & Security UK Limited, Bath, UK. John was the Chief Technologist for the ETI-funded Vessel Technology Assessment System project. John leads studies into marine power and propulsion, air-conditioning and hydraulics within the company and has published several papers on these subjects. Since 1999, John has been involved in a range of studies to assess the utility of EST for the commercial and military marine.

David Pearson holds the position of Senior Engineer at BMT Defence and Security Ltd. His primary domain is marine consultancy and auxiliaries design, with a focus on new and innovative EST. His previous experience includes modelling wind-based technologies for the ETI, among a range of other EST projects.

Beth Storey is an engineering placement student (June 2018 to July 2019) at BMT Defence and Security UK Ltd. She has recent experience modelling EST on commercial vessels including wind-based propulsion devices and hydrodynamic efficiency devices. Beth is a CFD specialist with air-domain modelling.

OPTIMAL ROUTING OF A WIND-POWERED CARGO VESSEL USING ENSEMBLE WEATHER FORECAST DATA

G. D. S. Davies and **T. Waseda**, The University of Tokyo, Japan

SUMMARY

This paper presents a novel optimal routing method using Numerical Weather Prediction wind-wave fields, created in conjunction with the Wind Challenger Project, a Joint Industry Project developing a partially wind powered cargo vessel. The devised algorithm optimizes way-points continuously to form a route that minimizes fuel consumption under a time constraint using the interior point method. The new Ensemble Averaging Method (EAM) takes forecast uncertainty into account using the ECMWF ENS-WAM ensemble data set and Monte-Carlo sampling. The EAM is evaluated against the optimal route found using the control forecast (CFR), considered the most accurate single forecast available. Simulations, run across 2016 on a trans-Pacific route, show that the EAM significantly outperforms the CFR in finding the analysis optimal route. In addition, when comparing using analysis data, both the mean engine power and mean error in arrival time were reduced by 16 BHP and 0.23 hours respectively.

NOMENCLATURE

BHP	Brake HorsePower (Ps)
ECMWF	European Centre for Medium-Range Weather Forecasts
EPP	Energy Prediction Program
JIP	Joint Industry Project
NCEP	National Centers for Environmental Prediction
NWP	Numerical Weather Prediction
SER	Saved Energy Ratio
WCP	Wind Challenger Project

1. INTRODUCTION

The WCP is a JIP that has developed designs for massive 10x50m hard-wing sails for use on cargo vessels to provide wind propulsion power alongside an auxiliary engine to reduce fuel consumption. This paper considers the routing of the bulk carrier design of the WCP which incorporates four of these sails [1] across the trans-basin route of Yokohama to Seattle.

Weather routing is constrained by wind-wave parameters and good quality forecasts allowing ship captains to select the optimal route, for which there are a multitude of tools and techniques available. The objectives of these routing methods are comprised of a combination of sub-objectives: passenger comfort, ship and cargo safety, fuel consumption, arrival time and distance traveled. For cargo ships a balance of these sub- objectives is necessary but it is important to note that a constraint on voyage time is key. Every moment spent at rest represents a loss for the shipping company,

With the reintroduction of wind power in the 21st century, a constraint on voyage time must be considered in the optimization process. The WCP could in theory be powered entirely by wind, however voyage times would become too long and unpredictable for the modern economy. Weather routing algorithms must make use of wind power in a reliable manner, to provide fuel savings while meeting shipping deadlines. This paper presents a routing method that makes use of ensemble forecast data to meet these requirements.

2. LITERATURE REVIEW

2.1 OPTIMAL ROUTING OF SHIPS

The first instance of a weather routing program was the development of the Isochrone method by James [2]. This served as the basis for many future developments in the field of optimal routing and its use has been independently studied and its benefits to improve the fuel consumption of ships confirmed [Nishiyama 3]. In addition to the Isochrone method, the multi-objective nature of ship routing has been recognized and techniques for this have been employed, using a wide variety of problem discretization and optimization methods. Walther [4] has published a comprehensive literature review on weather routing methods and the real world benefits of weather routing have been confirmed by captains on shipping routes in the Mediterranean [Delitala 5]. Despite the large amount of literature in this field, there has been a limited effort to include uncertainty into the optimization procedure which has been recognized as an area for further work by most researchers.

2.1 ROUTING OF SHIPS UNDER UNCERTAINTY

It has been recognized that the uncertainties surrounding the calculation of an optimal route have significant effects on the outcome. In the initial development of the modified Isochrone method Hagiwara [6] studied the stochastic optimization of wind assisted ships by estimating the correlation function of the forecast to the actual winds with which confidence intervals on the arrival time of the ship may be calculated. To improve on the reliability of shipping routes however, novel optimization techniques are required that can solve the multi objective problem of both minimizing fuel consumption and improving reliability in voyage time.

Use of ensemble weather data is currently the best method available to deal with the errors in forecasting. Work by Hoffschildt [7] looked at four methods to make use of the ensemble forecast and concluded that the mean route of the optimal route of all ensemble members has the most potential for safety and fuel consumption but that further work was required to reliably offer an improvement when compared to the analysis data. This method has been partially verified by Kristensen [8]. Work by Saetra [9] showed that the spread in optimal ship routes calculated for each ensemble member was related to the spread in the spread of significant wave height in the ensemble forecast. Ensemble data has also been considered in route optimization as a method to improve reliability in terms of constraint violation through the analysis of a single route in all ensemble members [Hinnenthal 10, Skoglund 11]. Preliminary work was performed in this area and while a comparison with the analysis data indicated a potential improvement in route reliability the researchers have concluded that further work be done to verify these findings. In addition there are many other papers that indicate that the use of an ensemble forecast may be better than the control forecast for weather routing.

3. METHODOLOGY

3.1 ROUTING ALGORITHM DEVELOPMENT

In order to implement an algorithm that can make use of the ensemble data a deterministic algorithm was developed. The routing method was chosen to be the optimization of the position of waypoints, from which a route can be constructed by interpolating between the waypoints using the Piecewise Cubic Hermite Interpolating Polynomial (PCHIP). This method has the benefit of covering an extremely large search space and can be easily modified for multi-objective optimization. The routing algorithm was coded in Matlab and for this paper constraints on arrival time, ship BHP, wind speed, wave height and avoiding sailing trajectories over land are considered.

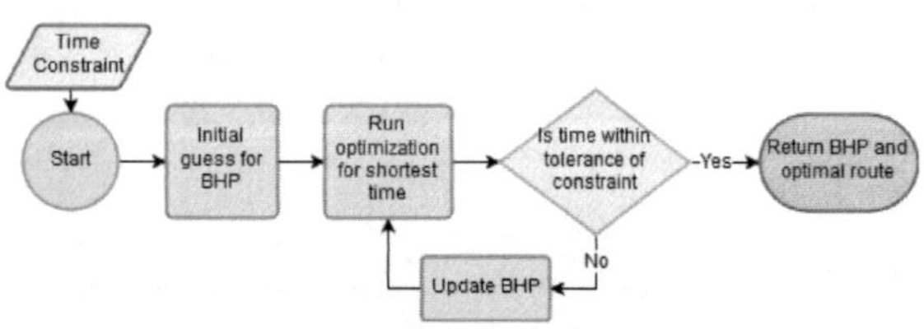

Figure 1. Overall structure of routing algorithm

The basic structure of the algorithm shown in Figure 1 relies on the sub-optimization problem of selecting the route with the shortest time at a constant BHP, hence maximising the contribution of wind energy. The BHP is then updated by a root finding algorithm using the bisection, secant, and inverse quadratic interpolation methods to select the BHP in which the vessel meets the time constraint.

For the time minimization algorithm, the interior point algorithm was selected for its suitability to problem scale, performance in terms of computational time and necessity to meet constraints at every step. The objective function is the arrival time of the ship along a specific route with a given BHP and wind/wave field. This function makes use of the ship hydrodynamics to calculate the time of voyage for the ship and is shown in Algorithm 1.

```
procedure ArrivalTime(X, Y, U, V, HS, MWP, MWD, BHP)
    CurrentTime ← 0
    Pathpoints ← Interpolate X, Y
    N ← length(Pathpoints)
    for i = 1 to N − 1 do
        Calculate distance and heading of Pathpoints(i) to Pathpoints(i + 1)
        Speed ← EPP(Pathpoints(i), heading, U, V, HS, MWP, MWD, BHP)
        TimeStep ← distance/speed
        CurrentTime ← CurrentTime + TimeStep
    return CurrentTime
```

Algorithm 1. Arrival time calculator

The WCP Bulk Carrier velocity as a function of engine power, true Wind Speed and Direction, and significant wave height, direction and period have been precomputed and are used by the routing algorithm as a lookup table. Algorithm 2 shows how the ship speed is calculated based on the lookup table and local weather conditions.

```
procedure EPP(x, y, heading, U, V, HS, MWP, MWD, BHP)
    u, v ← Interpolate U, V at (x, y)
    TWS ← √(u² + v²)
    TWD ← AngleFromNorth(u, v)
    TWA ← AngleBetween(TWD, heading)
    ShipSpeed ← Interpolate EPP Table at (TWA, TWS, BHP)
    if WaveDrag = True then
        hs, mwp, mwd ← Interpolate HS, MWP, MWD at (x, y)
        WaveAngle ← AngleBetween(mwd, heading)
        SurgeForce ← Interpolate Surge Force Table at (hs, mwp, WaveAngle)
        EHP ← Interpolate EHP table (ShipSpeed)
        Force ← EHP / ShipSpeed
        dEHP ← SurgeForce / Force
        ShipSpeed ← Interpolate EPP Table at (TWA, TWS, BHP, dEHP)
    return ShipSpeed
```

Algorithm 2. EPP

The interpolation function returns a set of evenly distributed points along the curve, and the distance between points was made to be less than the resolution of the meteorological data grid to ensure accuracy. For the trans-basin route 5 way points were found to be sufficient in mapping all feasible routes, but this number must be updated depending on route.

The interior point algorithm optimizes the position of a set of waypoints at a constant BHP to minimize the arrival time along the route by estimating the gradient of the objective function in each direction from a set of initial points and iterating until a solution is found. This method can be costly in terms of computational time but can be improved by calculating the gradients in parallel and using the last found optimal route as the starting position in subsequent calls to the algorithm.

To verify the performance of the developed algorithm the output was compared to the results of the Isochrone method, a long used and trusted routing method [Nishiyama 3]. Comparison showed that the developed algorithm matched the route found by the Isochrone method in a shorter computational time. Additionally, the optimal BHP of the route was on average lower as routes were smoother than the sometimes jagged Isochrone route.

3.2 UNCERTAINTY QUANTIFICATION

For this paper the uncertainty in weather forecasts is not directly quantified and instead, treating each ensemble member as a sample taken from the distribution of future events, a Monte Carlo based method is used to account for uncertainty. With only 50 members, statistically this is not enough for a full Monte Carlo simulation, but it has the advantage that each member of the ensemble is coherent in space and time and hence represents a future that is physically possible. In addition Monte Carlo samples should be randomly selected from the distribution, while the members of the ensemble forecast are non-random forecasts that are calculated from perturbed values of the best available analysis data at t=0. Nevertheless this is a suitable approximation to make use of ensemble data [12].

4. TESTING

4.1 METEOROLOGICAL DATA SETS

For routing under uncertainty the ECMWF ENS-WAM wind-wave dataset was used because it makes use of a coupled atmospheric-wave model for the ensemble forecast. From the ECMWF three types of data are used in this paper; Analysis, Control Forecast and Ensemble. Analysis is considered fully accurate and used to measure real world routing performance. Control Forecast data is considered the best single estimate available for future weather patterns. Ensemble forecasts are a set of 50 forecasts with perturbations applied at the start of the forecast to give the widest range of energy states to the model and hence aim to represent the full range of possible future outcomes [13].

4.2 SIMULATIONS

A series of simulations were run over the year of 2016 with one simulation run of a 13 day voyage each month. This required the optimization algorithm to be executed 52 times; for 50 ensemble members, the control forecast and the analysis data each month. These simulations were run in the absence of wave forcing due to a lack of data for a 4 wing vessel. Wind and wave forcing data for a 1 wing vessel was used to verify the correct operation of the algorithm, but due to minimal forcing from the wind the results deviate only slightly from the Great Circle route and were not suitable for a thorough comparison of methods. The mean ensemble route is calculated by finding the geographical mean of the optimal waypoints from each ensemble member.

4.3 RESULTS

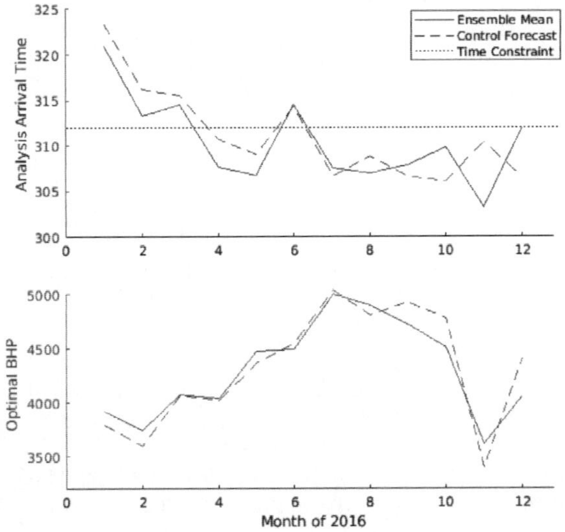

Figure 2: Optimal BHP and Arrival time in Analysis data over the year 2016 for ensemble mean and control forecast routes

Figure 2 shows the variation in optimal BHP and arrival time in hours for the ensemble mean (EAM) and control forecast (CFR) route in the analysis data. Both methods calculate a similar optimal BHP throughout the year with the exception of months 9, 10 and 12 when the ensemble mean significantly improves on the control forecast value. Moreover, the ensemble mean route for these months has a lower error on the arrival time and generally throughout the year has a lower mean error of 4.15 hours compared to the control forecast mean error of 4.38 hours on top of an average improvement in the optimal BHP of 16.0.

In a mathematically rigorous Monte Carlo simulation it would be expected that the mean of the samples would tend towards the true outcome represented by the analysis forecast. The EAM was therefore anticipated to be a better approximation of the analysis route on average than the CFR alone and so the error in terms of distance between the routes should reduce for the EAM. To quantify this, an error parameter was devised that integrated the distance between the routes along their lengths. This was then normalized against the length of the analysis route to create a measure: "relative error to analysis route". If a route perfectly matches the analysis route the value of this parameter will be 0.

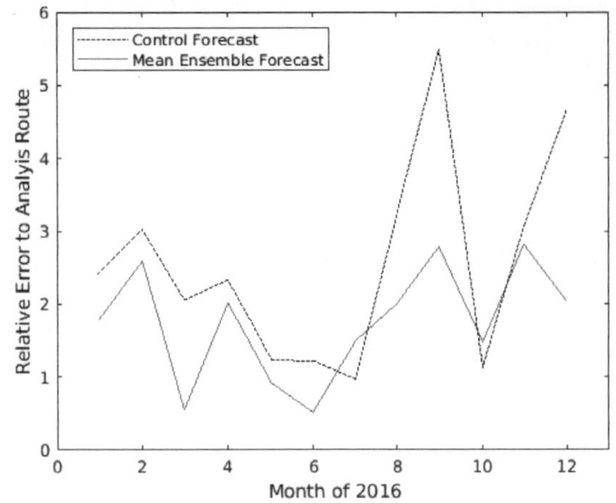

Figure 3: Relative error to analysis route over the year 2016 for ensemble mean and control forecast routes

Figure 3 shows that the mean ensemble forecast consistently outperforms the control forecast at matching the analysis route and the spread of the error is reduced. Across the year the mean error of the control forecast was 2.70 while the mean ensemble error was 1.75, a 35% improvement. Figure 5 shows a plot of the routes from the 3rd month, March, showing that the EAM almost exactly matches the analysis optimal route, while the forecast optimal significantly deviates from it.

How the relative error to analysis route changes with number of ensemble members used was analyzed to check the rate of convergence shown in Figure 4. Values across the entire year were used and averaged to give representative values of the errors. The bold line shows the mean rate of convergence. On average using roughly 25 ensemble members is sufficient to find a route that best matches the analysis.

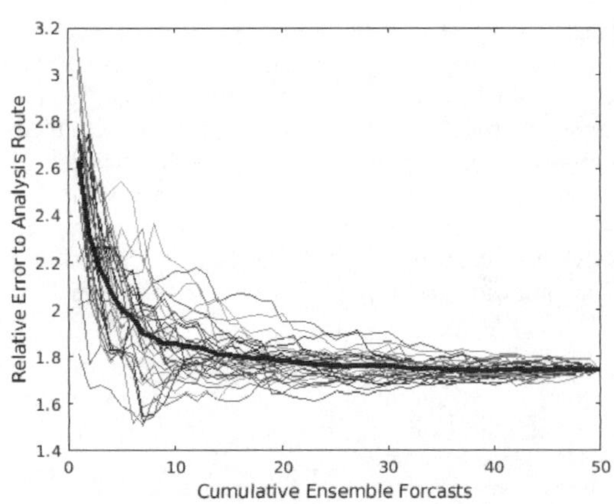

Figure 4: Relative error to analysis route against number of ensemble members used in route calculation

5. DISCUSSION

Use of the EAM shows an improvement over the CFR in terms of constraint violation, efficiency and most significantly in its ability to seek the analysis route. It is clear this method has the potential to create better routes than the current deterministic methods. However, these benefits are modest and require substantially more work to verify statistically. The EAM has been shown to consistently outperform the use of the CFR but at the expense of longer computational times. By studying the convergence of this, it is found that 25 members are sufficient to allow the computational time for selecting the ensemble mean route to be halved in the case of the ECMWF forecast. This will benefit captains when searching for an optimal route in real time.

The 7th month, July, was the only month of the year for which the CFR significantly outperformed the EAM in terms of error to the analysis route shown in figure 6. It is known that trans pacific winds are weakest during the summer period and in this case all 3 routes are closely grouped near to the great circle route indicating a reduced effect from wind forcing. While the relative error of the control forecast may be lower, in absolute terms both routes had relatively low errors to the analysis optimum.

One of the key issues of weather routing is how to incorporate incoming forecasts while asail. Most routing algorithms consider the route calculated at time 0 for the entire route, but in reality, ECMWF ENS-WAM forecasts are provided every 12 hours and so routing can be updated in real time. Figure 7 shows an example of routing when updating at every 12 hours on the trans-basin route based on the control forecast. As it continues, the ship follows the course in red, while the green represents the global optimum in analysis data. Finding the global optimum of the stochastic optimization problem is very complex and it may not be possible to find the optimum apriori, as calculated using analysis data. Therefore a route that can better seek the analysis optimal route from the start may have significant effects when compounded across the length of the route.

5.2 FURTHER WORK

To develop this method further work must be put into 3 areas; improving the accuracy and performance of the optimization algorithm, conducting large scale simulations on historical data, using other data sets such as those offered by the NCEP for comparison, and a more quantitative approach to validate the improvements of the EAM statistically.

The EAM also has room for improvement such as the estimation of the PDF of uncertainty in weather parameters to allow a full scale Monte Carlo simulation to be performed. While this paper only considers uncertainty in the weather forecast, it is important to note that uncertainty manifests itself in many parts of the weather routing optimization problem. As a long term goal these must be quantified as, while individually they are insignificant, their interaction through a complex non-linear model can become difficult to predict.

Most importantly the EAM requires extensive testing to verify the performance on different routes around the world and across a larger time period. Furthermore in order to validate the findings of this research these methods must be tested on a real life ship. Measurements should be made from the ship to verify that the weather forcing inputs are accurate enough to discover an optimal route. With the installation of a trial sail this validation can come soon, however with a single sail the forcing effects will be small and may be difficult to isolate from the engine power.

Additionally, thought must be given to how the code can be improved once the Wind Challenger Project is operational by using data collected on actual voyage and local wind speed measurements aboard the ship. This should extend to how the code can use real time information while any wind powered ship is at sea in the real world. For example, how well can measured parameters from the ship and forecast updates be used to improve the routing?

6. CONCLUSION

This paper has developed a novel routing method for the Wind Challenger Project (WCP) using an ensemble forecast. While showing modest improvements on the control forecast for meeting the arrival time constraint and reducing BHP it has shown to be most effective at finding the analysis optimal route. If compounded over the course of a route with continual forecast updates, this has the potential to hugely improve the operation of the WCP in terms of reliability and fuel consumption. Based on this initial research a more thorough study must now be conducted to quantify the scope of this improvement.

7. ACKNOWLEDGEMENTS

I would like to thank Dr Takuji Waseda for his introduction to the field of weather routing and for his continual support, Ouchi Ocean Consultancy and Mitsui O.S.K. Lines for their support and funding to attend this conference, and Dr Matthew Eaton for supervising the Masters thesis research from which part of this work is based.

8. REFERENCES

1. OUCHI, K., "Wind Challenger the Next Generation Hybrid Sailing Vessel", *Third International Symposium on Marine Propulsors*, 2013.

2 JAMES, R. W., "Application of wave forecast to marine navigation", *Technical report Naval Oceanographic Office*, 1957.

3. NISHIYAMA, H., "A Study on Ocean Passage Route using Weather Routing", *日本航海学会第 124 回講演会にて講演 Japan Institute of Navigation*, 2011.

4. WALTHER, L, "Modeling and Optimization Algorithms in Ship Weather Routing", *International Journal of e-Navigation and Maritime Economy*, 2016.

5. DELITALA, A.M.S., "Weather routing in long-distance Mediterranean routes", *Theoretical and Applied Climatology*, 2010.

6. HAGIWARA, H., "Weather Routing of (Sail-Assisted) Motor Vessels, *PhD Thesis TUDelft*, 1989.

7. HOFFSCHILDT, M., "Potential Benefit of Ensemble Forecasts for Ship Routing", *Technical report, ECMWF*, 1999.

8 KRISTENSEN, N. M., "Weather routing: Sensitivity to ensemble wind and current input", *University of Oslo Master thesis in Geosciences Meteorology and Oceanography*, 2010.

9. SAETRA, Ø., "Ensemble Shiprouting", *Technical report ECMWF*, 2004.

10. HINNENTHAL, J., "Robust Pareto-optimum routing of ships utilising deterministic and ensemble weather forecasts", *Ships and Offshore Structures*, 2010.

11. SKOGLUND, L., "A comparative study of deterministic and ensemble weather forecasts for weather routing", *Journal of Marine Science and Technology (Japan)*, 2015.

12. PALMER, T. N., "Predicting Uncertainty in Numerical Weather Forecasts", *International Geophysics*, 2002.

13. BUIZZA, R., "The new ECMWF Variable Resolution Ensemble Prediction System (VAREPS): methodology and validation", *ECMWF*, 2006.

9. AUTHOR'S BIOGRAPHY

G D S Davies currently works at the design consultancy Takram and at the University of Tokyo Design Lab, and is a recipient of a Scholarship from the Daiwa Anglo-Japanese Foundation. He holds a Masters in Mechanical Engineering at Imperial College London and has worked at the University of Tokyo as a project researcher under the supervision of Dr Takuji Waseda as part of the Wind Challenger Project.

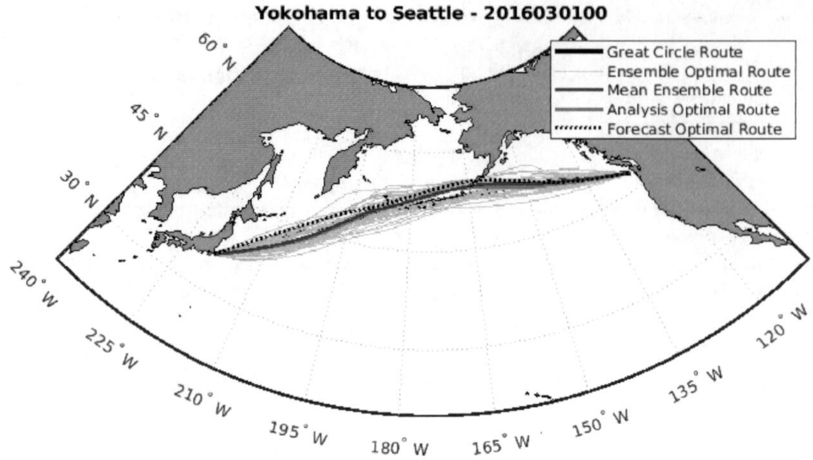

Figure 5. Routing in March

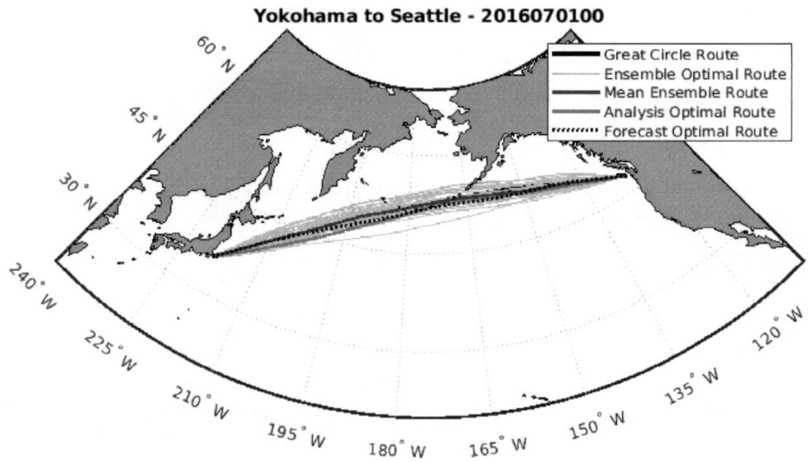

Figure 6. Routing in July

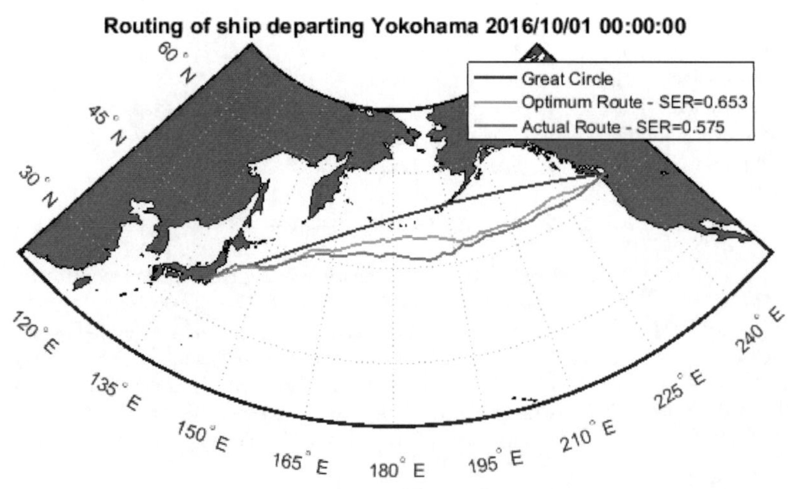

Figure 7. Trans-basin routing showing the optimal route in analysis data and the actual route taken by updating the heading with every new control forecast

SEAKEEPING AND MANOEUVRING FOR WIND ASSISTED SHIPS

R. Eggers and **A. S. Kisjes,** Maritime Research Institute Netherlands, the Netherlands

SUMMARY

Research on Wind Assisted Ship Propulsion (WASP) has until now focussed on performance in steady conditions whereas unsteady behaviour is an important part of ship behaviour. For "conventional" manoeuvring and seakeeping, performing direct "free running" scale model tests has been successful to gain understanding and to improve ship designs. In the "WindLab" project the opportunity was taken to extend that approach to aerodynamics: modelling real wind with a simplified wind tunnel in the basin. A unique set-up was tested for its suitability and first data was collected on the performance of wind assisted ships in unsteady scenarios.

Figure 1: "MARIN Hybrid Transition Coaster" with Dynarigs (left) and Flettner rotors in the "WindLab" test set-up

NOMENCLATURE

[Symbol]	[Definition] [(Unit)]
β	Leeway (drift) angle (deg)
δ	Steering angle (deg)
ψ	Yaw (heading) angle (deg)
AWA	Apparent Wind Angle (deg)
AWS	Apparent Wind Speed (kn)
C_L, C_D	Lift and Drag Coefficient [-]
C_X, C_Y, C_K, C_N	Coefficients for forces in ship fixed x- and y-direction and moments in x- and z- direction [-]
TWA	True Wind Angle (deg)
TWS	True Wind Speed (kn)

1. INTRODUCTION

At the core of MARIN's mission is to make ships better in collaboration with industry. Reducing CO_2 emissions, down to "zero emission", is part of that mission, following the IMO ambition for 2050. Many methods are available: general optimisations, slow steaming and batteries or alternative fuels as energy carriers are probably unavoidable. Especially the change in energy carriers will likely result in larger expenses for energy and shorter range (the latter because of lower energy density in the alternatives for diesel). In this context Wind Assisted Ship Propulsion (WASP) may be seen as a method to save operational costs and to increase the range (on average) on routes with favourable wind. It can of course also be used on its own, as is mostly done in current market introductions. MARIN aims to provide its contribution to help shipping make best use of Wind Assisted Ship Propulsion. To this end MARIN conducts research that is openly published, organizes Joint Industry Projects and directly helps companies with their own projects.

As described in the following sections, MARIN was already active in predicting the emission (and fuel) saving of WASP vessels, and providing recommendations for improvement in steady conditions. Results from published research are given in e.g. [1] and [2].

A recently available fund for background research allowed to extend the field to dynamics in the "WindLab" project.

1.1 STATE OF RESEARCH

After earlier short periods (e.g. in the 1980's), Wind Assisted Ship Propulsion is firmly back on the agenda of research for ship performance. At present, the driving factors are believed to be the potential for reducing operational cost in the context of high fuel prices and the reduction of ship air emissions in view of climate change and upcoming more stringent regulations.

Nevertheless, both market introductions and research have still been modest when comparing it against other fields of ship propulsion, e.g. Energy Saving Devices. Installations on commercial cargo vessels have been restricted to a few type of devices: Flettner rotors, retractable rigid wings and (temporarily installed) suction wings. All of them have been dimensioned for relatively modest fuel/CO_2 savings (lower than about 15%). Earlier research showed that

further design development and when owners/operators accept some change in operations, then much higher savings can be accessed using Wind Assisted Ship Propulsion.
As one would expect, the research has focussed on performance in steady conditions. A broad overview of expected results is given in [3]. The impact of wind assistance on heel, drift, steering angle and the best actuator/sail settings feature in most publications. MARIN also contributes in this field. E.g., the influence of main particulars on the savings with WASP was demonstrated [4]. MARIN, and others, are continuing in this field to make sure that ship designs will be specifically tailored for WASP.

However, ships hardly sail in steady conditions. Conditions in real operations continuously vary and ships need to respond to them.

1.2 UNSTEADY PERFORMANCE SCENARIOS

Ships mostly sail in environmental conditions that are dynamic: wind is never really constant in magnitude and direction, and waves will always be present at sea, leading to ship motions. Furthermore, the desired behaviour of the ship itself may be unsteady, for example when the ship is accelerating, performing a (crash-) stop manoeuvre or changing course. Requirements on unsteady behaviour can be related to the performance to use minimal energy for a certain speed or it can be related to regulations that the ship must conform to. Within WindLab, the following unsteady scenarios were considered relevant for wind assisted ships:

1 Manoeuvring: Zig-zag, turning circle and crash stop – The wind assistance results in a side force and yawing moment and a change in flow over the rudder that affect the performance in these manoeuvres as mandated by IMO.

2 Course keeping in stern quartering waves and wind – Stern quartering wind is very favourable such that potentially the propeller can be throttled down. However, that in turn creates very little flow over the rudder precisely in a condition that is demanding in terms of course keeping.

3 Sailing in head or bow quartering wind and waves. In these conditions, the performance of a WASP ship may in fact be less than a conventionally driven vessel due to larger air drag. One could therefore ask if the IMO minimum power requirement will effectively determine the installed power on a WASP ship.

The scenarios above can be judged just by themselves. However, failing to meet established thresholds may in fact also lead to lower overall savings if e.g. the acceptable wind propulsion needs to be limited in certain conditions.

1.3 WINDLAB PROJECT

Unsteady behaviour could be modelled directly with calculations of varying fidelity. However, complete validated force descriptions with WASP are not directly available including dynamics. At this moment, calculations either rely assumptions with empiricism or need to go in depth with e.g. CFD. At MARIN it is believed that in manoeuvring and seakeeping, direct experiments, with the ship in its full environment and sailing by itself is both practical and trustworthy as the full physics happens by itself, without additional assumptions (but including potential scaling effect). In the present case it is required then to also model the aerodynamics.

Previously, MARIN has conducted tests where aerodynamic forces were obtained from a database and then applied in the tests with dynamically controlled winches and lines to the ship model, see [3]. However, this approach still makes assumptions. For instance, the description of aerodynamic forces does not include all dynamics.

In the present project the main objective was to directly model the aerodynamics using "real wind". That requires a special experimental set-up modelling both aero and hydrodynamics for a sailing ship at the same time. The authors are not aware of an existing set-up with these capabilities. Thus, developing such a set-up was judged to be research by itself.

The following research questions were specified:

- How can aero- and hydrodynamics be modelled at the same time in one test-setup and what is the quality of tests in this test set-up?
- As a first indication: what is the performance of a ship with WASP in unsteady conditions?

2. TEST SET-UP

A simplified wind tunnel was built within MARIN's Seakeeping and Manoeuvring Basin. Figure 2 through Figure 4 show the final arrangement. It comprised:

- A "blowing" array 3x8 wind fans blowing wind into the test section (green in Figure 2)
- An "aspiration" array of 2x5 wind fans extracting air from the test section (red in Figure 2)
- A sheet of transparent sail cloth closing one side of the test section, from the blowing array to the aspiration array (orange in Figure 2)
- A sheet of transparent sail cloth closing the top side of the test section

This set-up was selected after trying various configurations of sheets to contain the flow at smaller scale. The configured sheets of sail cloth were found to be necessary to enforce a local wind field that is governed by the wind fans, not the surrounding conditions around the basin carriage. One side of the test section was open. This was required to have access to the ship model. The "aspiration" array of wind fans was found to be necessary to make sure that wind travels the full length of the test section; to prevent early recirculation. Honeycomb material was fitted in the outflow of the blowing wind fan array, to have a better alignment of the flow in axial direction.

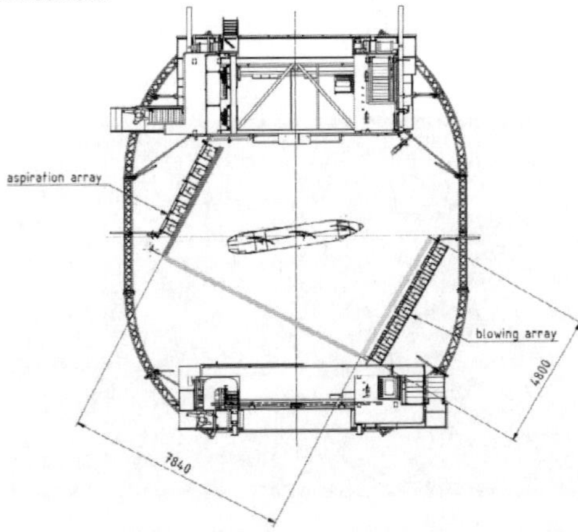

Figure 2: Top view of test section

Figure 3: Blowing array of wind fans with honeycomb

Figure 4: Aspiration array of wind fans

3. MODELLING OF WIND

When a ship is using wind propulsors as additional to its main propulsion, the ship experiences the "apparent wind", which is a combination of the "true wind" and its own ship speed. The apparent wind speed (AWS) and the apparent wind angle (AWA) can be related to the true wind speed (TWS) and true wind angle (TWA) by accounting for the ship's speed, heading and leeway. The TWA is the angle of the true wind relative to ship's track, whereas the apparent wind is the experienced wind, relative to the ship's heading. The apparent wind speed, AWS, is calculated applying the formula as shown in equation (1):

$$AWS = \sqrt{\begin{array}{l}(V_s \cdot \cos(\beta) + TWS \cdot \cos(TWA - \beta))^2 + \\ (V_s \cdot \sin(\beta) + TWS \cdot \sin(TWA - \beta))^2\end{array}} \quad (1)$$

Similarly, the apparent wind angle, AWA, is defined as given in equation (2):

$$AWA = \arctan\left(\frac{V_s \cdot \sin(\beta) + TWS \cdot \sin(TWA - \beta)}{V_s \cdot \cos(\beta) + TWS \cdot \cos(TWA - \beta)}\right) \quad (2)$$

According to the definition for specifying the wind in equation (1) and (2), this means that:

- 0° is wind from bow;
- 90° is wind from port side;
- -90 or 270° is wind from starboard;

For the calculation of force coefficients it is needed to have the apparent wind speed and direction. It is assumed that in the analysis of the tests:

- the AWS is constant;
- the AWA is a combination of the angle of the wind array with respect to the carriage, ψ_{SetUp}, the direction of the measured wind, ψ_{wind} and the heading of the ship, ψ according to:

$$AWA = \psi_{SetUp} + \psi_{wind} + \psi \quad (3)$$

- the TWS and TWA are reversely calculated based on the AWS and AWA.

It is noted that the experiment thus enforced a situation that is different than for a real ship. In the experiments the apparent wind angle (relative to the carriage) and speed were fixed, whereas in reality the true wind angle (relative to a straight track) and wind speed are fixed.

4. SHIP DESCRIPTION

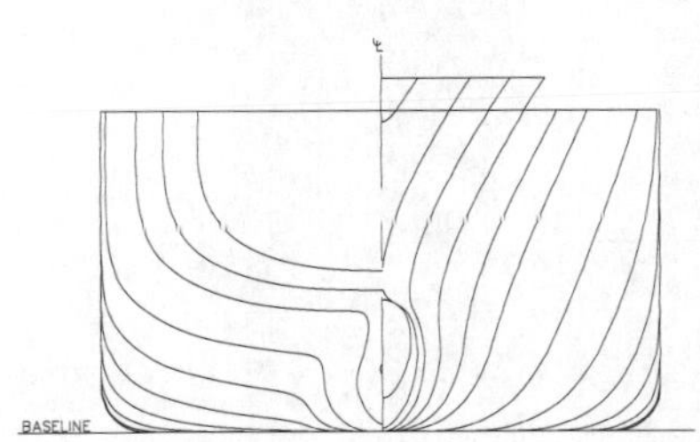

Figure 5: Body plan "MARIN Hybrid Transition Coaster"

An existing ship model was used, for which all data is free to publish. The hull was originally designed for a 99548 ton LNG carrier. In the present case it is considered to be a 4146 ton coaster. The smaller scale ratio helps to not have excessively low wind speeds at model scale. Figure 5 shows the body plan and Table 1 the main particulars of the ship.

The ship was equipped with a single rudder. The hull was not specifically designed for WASP, therefore the capability te develop side force was limited and the centre of effort of side force was predicted to be located further forward than would ideally be desired for wind propulsion. It was therefore decided to add additional "skegs" approximately next to the propeller, both on port side and starboard.

In reality, the GM of a vessel with Dynarigs is expected to be lower than with Flettner rotors if the design is otherwise identical. That is opposite to what was obtained in the tests, which was dictated by the measurement and control instrumentation on board. Nevertheless, data sourced from developers of these devices suggest that the real differences in GM are not large with the dimensions as considered in this project. Flettner rotors also need a substantial support structure and foundation.

Table 1: Main particulars of "MARIN Hybrid Transition Coaster"

Length between perpendiculars	L_{PP}	68.7	m
Breadth moulded on WL	B	14.0	m
Draught	T	4.4	m
Displacement mass in seawater	Δ	4146.0	t
Transverse metacentric height Dynarigs	GM	0.69	m
Transverse metacentric height Flettner rotors	GM	0.67	m
Block coefficient	C_B	0.76	-

The model was sequentially equipped with Dynarigs (see Figure 6) and Flettner rotors (see Figure 7). The principal dimensions are given in Table 2 and Table 3.

Figure 6: Model with Dynarigs

Table 2: Main particulars of three identical Dynarigs

Span (incl. yards)	s	29.9	m
Average chord	c	16.4	m
Combined projected area	A	1474.5	m^2
Overall air draught from WL	h_{air}	38.4	m

Figure 7: Model during tests with Flettner rotors

Table 3: Main particulars of three identical Flettner rotors

Height	h	18.0	m
Diameter cylinder (disk)	d	3.0 (6.0)	m
Combined projected area	A	162.0	m^2
Overall air draught from WL	h_{air}	25.56	m

5. TEST, MEASUREMENT AND ANALYSIS PROCEDURES

The tests were conducted according to MARIN's normal methods for free running tests. Thus, the model was sailing completely free, propelled by the wind propulsors and the propeller and steered by the rudder on auto pilot control. Thin lines were attached at the bow and stern for security and to accelerate and stop the model. They were slack during the actual tests. Positions were measured optically. The wind propulsors were mounted on force measurement frames.

The measurement for each test was always started at rest, so that any drift in sensors readings could be measured and compensated for.

Heeling and oscillation tests were conducted to determine the inertia properties of the vessel and the wind propulsors. The measured wind propulsor forces as shown in this publication were corrected for inertia effects (including static heel) and are presented in the plane parallel to the water surface. The presented forces are converted to full scale using Froude scaling only.

6. TEST PROGRAM

The following test series were conducted:

- Wind quality tests
- Tests with Dynarigs in close hauled (bow quartering) wind conditions:
  - Speed runs and roll decay tests
  - Zig-zag tests
  - Seakeeping tests in regular and irregular bow quartering waves
- Tests with Flettner rotors in broad reach (stern quartering) wind conditions:
  - Speed runs and roll decay tests
  - Zig-zag and crash-stop tests
  - Seakeeping tests in regular and irregular stern quartering waves

A selection of these tests is described in this publication.

7. RESULTS

7.1 WIND FIELD

Prior to testing the ship model with wind propulsors, the wind field was first calibrated and measured without the model being present. Figure 8 shows the wind speed and angle relative to the test section at three longitudinal positions in the test section as used in bow quartering wind with Dynarigs. The settings for this condition were found after some tuning with the rotation speeds of the wind fans. It is judged that, although not perfect, the profile was rather constant over height and longitudinal position. However, some shear is present in the wind field. The wind angle, ψ_{wind}, varied between about -2 and 7 deg depending on height and longitudinal position. The standard deviation was on average 7% for the wind speed and 3 deg for the wind angle.

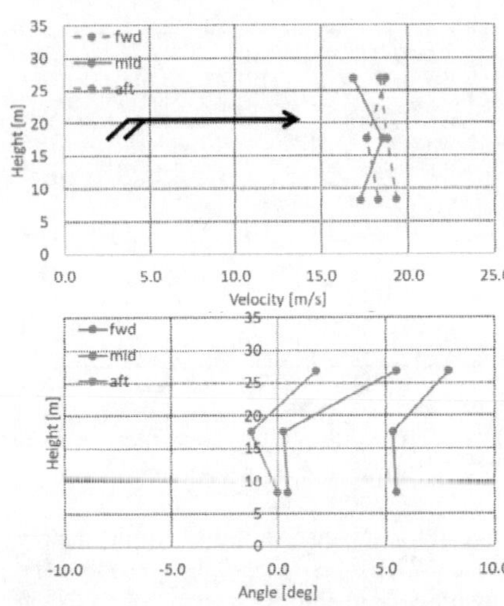

Figure 8: Profiles for wind speed (above) and wind angle (below)

The wind quality in broad reaching wind with the Flettner rotor was worse. The speed and angle varied much more as function of position and time. This was due to the "driving wind" wind of the carriage that apparently cannot be eliminated completely.

7.2 EQUILIBRIUM CONDITIONS AND FORCES WITH DYNARIGS

7.2 (a) Force Balance

For a ship that partially or completely uses wind propulsors for its propulsion, the ship also needs to balance aerodynamic wind forces with the hydrodynamic forces of the ship to keep course. This means that there needs to be force equilibrium in transverse y-direction and moment equilibrium around the horizontal x-axis and vertical z-axis of the ship. To counter the aerodynamic forces, the ship will adopt a leeway (drift), rudder and heeling angle.

The equations of motion are given in equations (1) through (4) to indicate the dominant force and moment components that are present during course keeping with wind-assistance. In these equations, only the thrust of the propeller and the forces of the sails were measured. Figure 9 shows a representative equilibrium condition from the tests with Dynarigs. The adopted sign conventions can be derived (a negative sign is displayed for negative quantities).

$$T + F_{X,Sails} - R_0 - R_R - R_i - R_{SS} = 0 \quad (1)$$

$$F_{Y,Sails} + F_{Y,SS} + F_{Y,R} + F_{Y,H} = 0 \quad (2)$$

$$F_{Y,Tot,Sails} \cdot (KG - z_{CE}) + F_{Y,SS} \cdot (KG - z_{SS}) + F_{Y,R} \cdot (KG - z_R) + F_{Y,H} \cdot (KG - z_{CLR}) = 0 \quad (3)$$

$$F_{Y,Tot,Sails} \cdot x_{CE} + F_{Y,SS} \cdot x_{ss} + F_{Y,R} \cdot x_R + F_{Y,H} \cdot x_{CLR} = 0 \quad (4)$$

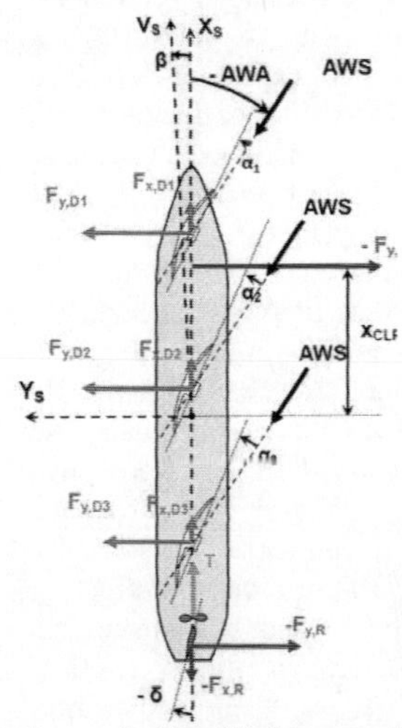

Figure 9: Equilibrium condition and sign condition (for sailing close hauled with Dynarigs)

7.2 (b) Speed runs in calm water

After establishing the best Dynarig trim angle, several tests were done for varying ship speed. The purpose of these runs was to compare the measured forces to known references and to establish a baseline for the manoeuvring and seakeeping tests to follow. Note that the TWA effectively varies as a function of ship speed. The equilibrium results are shown in Table 4.

To show the contribution of wind-assistance, two parameters are introduced: P_D/V_S^3 and the wind assist (WA) ratio. The power to cubic ship speed is judged to be a convenient parameter, because it allows to compare between similar ships speeds as in Figure 11 (although extrapolation should not be taken too far). The wind assist ratio compares the sail thrust to the total required thrust as in equation (5):

$$WA = \left(\frac{F_{X,sail,tot}}{F_{X,sail,tot} + T} \right) \cdot 100\% \qquad (5)$$

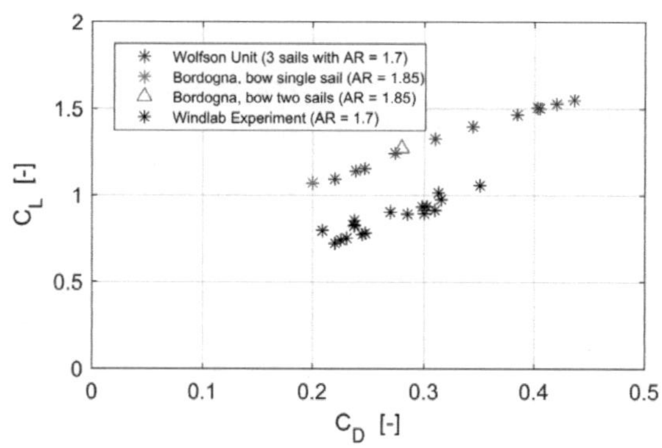

Figure 10: Comparison of lift and drag coefficients

In Figure 10 the obtained lift and drag coefficients in the experiments are compared to two other known references: data generated by the Wolfson Unit for the North Sea Region SAIL project [6], and a publication by Bordogna et. al [7]. It is firstly noted that lower lift coefficients were experienced in the present project. This was deliberately done, with relatively small angles of attack. The hull could not support a proportionally large side force as the hull was not designed for it. However, it can also be seen that the lift is low in relation to the drag compared to the other data. The reason for this is not yet clear. Separate CFD calculations and wind tunnel tests showed that scale effects can only explain a very small part of the gap. The other experiments shown in Figure 10 were also conducted at model scale, though at higher Reynolds numbers. The difference could have to do with variability of flow directions (see section 7.1) or blockage affecting the flow in the test set-up. It is judged that when using this test set-up to replicate accurate aerodynamic forces, then still some calibration is required using either the wind conditions or the wind propulsors sizing in order to match a known target for the wind propulsors forces.

The result with Flettner rotors was similar: also there it is seen that the observed force coefficients are lower than known from literature.

Table 4: Results for different ship speeds

Vs [kn]	12.1 (No wind)	7.9	11.7	12.6	13.1	13.2
δ_{ave} [deg]	-1.5	-25.3	-17.4	-15.7	-14.9	-14.6
β [deg]	-1.2	2.5	3.4	2.8	2.3	2.4
φ [deg]	0.3	-6.8	-10.3	-10.5	-10.6	-10.5
TWS [m/s]	-	15.1	13.8	13.5	13.4	13.3
TWA [deg]	-	-40	-52	-53	-54	-54
AWS [m/s]	6.2	18	18	18	18	18
AWA [deg]	0	-29	-33	-34	-34	-34
$\alpha_{Dyn,ave}$ [deg]	-	5.0	9.0	9.7	10.1	10.0
$C_{L,ave}$ [-]	-	0.42	0.80	0.82	0.86	0.83
$C_{D,ave}$ [-]	-	0.27	0.21	0.24	0.24	0.24
$F_{Y,Dyn,tot}$ [kN]	-3	146	232	240	245	245
$F_{X,Dyn,tot}$ [kN]	-8	32	70	72	77	76
T [kN]	180	95	121	150	172	185
WA [%]	-4.8	25	37	32	31	29
P_D/V_S^3 [ton/m]	6.19	8.09	4.13	4.44	4.74	5.07

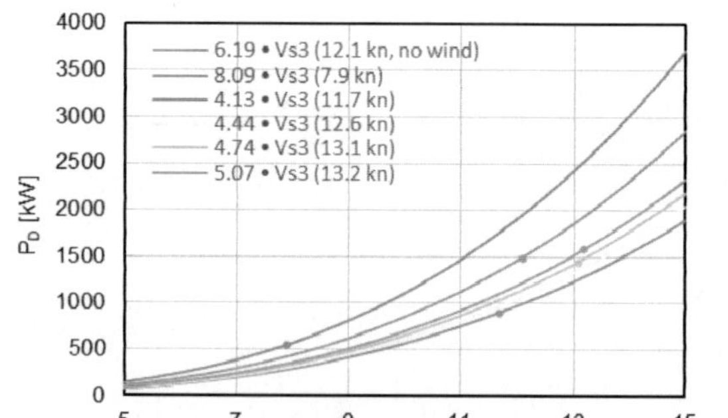

Figure 11: Delivered power of propeller versus ship speed (P_D/V_S^3 is shown in the legend)

Table 4 and figure 11 show that sailing slower leads to a higher wind-assist (WA) ratio and a lower power. The proportion of wind propulsion is larger, because the propeller thrust can decrease while the wind propulsion thrust remains high. However, there is a limitation. With decreasing propeller thrust and ship speed, the flow over the rudder reduces, making the rudder less effective. At the same time the side force and yaw moment do not decrease at the same rate, such that the rudder angle must increase to keep the vessel on course. The larger rudder angle leads to a large rudder resistance, which at a certain point is not sustainable anymore. At a ship speed of 7.9 kn, the ship is not capable to keep course. The rudder cannot generate sufficient side force. The ship speed needs to increase or the design needs to be adapted.

7.3 ZIG-ZAG TESTS AND SIMULATIONS WITH DYNARIGS

In general, zig-zag manoeuvres are preferably conducted in calm water without any external environmental forces, such as wind. In such test conditions, a (near) zero rudder angle is needed to sail on a straight course. However, the impact of wind propulsors should of course also be checked in wind from various angles. And then the aerodynamic force likely requires a steering angle, δ_0 to keep course. In the present tests it was found that the ship requires an average rudder angle of -15° (trailing edge to port side) for the chosen condition and speed. IMO in its resolution MSC 137(76) [8] does not make any statement what constitutes a "zero rudder angle". For the present tests the rudder angle required for a straight course has been selected as the reference angle (starting point) for the zig-zag test. Thus, the zig-zag "10°/10°" is conducted with rudder execute angles of -5° and -25° whereas normally it would be 10° and -10°. The yaw angle in the analysis is represented with respect to the centre line of the basin. In order to have a more relevant wind angle, the ship was sailing under a yaw angle of about 8.5 deg through the basin, which is seen as the zero level for yaw in the results.

After the execution of the zig-zag 10°/10° model tests, it was observed that the realised yaw check angles were slightly wrong with 2 to 3 degrees. This error has a slight effect on the overshoot angles and initial turning ability as shown in the simulations presented below.

7.3 (a) Tests

In Figure 12 the time traces of the yaw angle, rudder angle, the thrust and transverse forces of the Dynarigs are shown. Bird eye views, showing the tracks from above, are presented in Figure 13. The derived overshoot angles and initial turning ability are shown in **Error! Reference source not found.**. Experimental results are shown combined with simulation results. The simulations are discussed below.

When first considering the experiments ("Exp") it can be seen that when the manoeuvre is initiated to port side, the overshoot angle in the zig-zag is larger than when initiating to starboard. The wind forces introduce asymmetry. Various physical processes play a role. When initiating to port side, the apparent wind angle becomes larger, increasing the thrust such that the speed and momentum of the vessel increase. As a consequence the propeller thrust reduces as well as the flow over the rudder. This can explain a larger overshoot angle. However, at the same time, having the thrust further outboard (due to increasing heel) and the increased yawing moment from sails and hull would lead to a lower overshoot angle. Apparently the first effect is more important. The opposite argument can be made for the manoeuvre initiating to starboard. The differences in the second overshoot show the same trend, but they are larger.

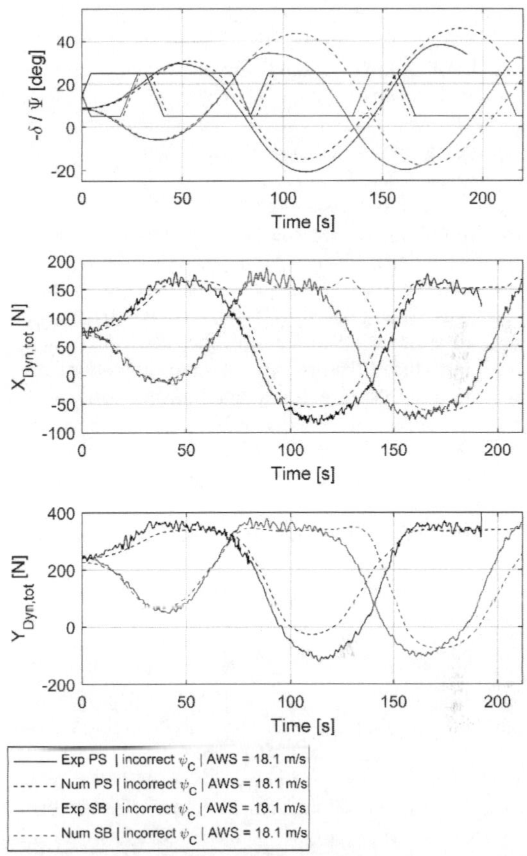

Figure 12: Time traces of zig-zag tests

Corresponding with the trend of the overshoot angle, the initial turning ability is larger for the zig-zag test initiating to port side when compared to starboard and becomes respectively $1.74 \cdot L_{pp}$ and $2.10 \cdot L_{pp}$.

7.3 (b) Simulations

Time domain simulation were used to investigate the influence of the incorrect implementation of the yaw checking angles on the manoeuvring characteristics. Also, the effect of forcing apparent rather than true wind was investigated as well as the behaviour of the ship without wind force.

Numerical simulation solve the equation of motion (see equation 1-4 in the time domain and require a mathematic

model that describes the forces of the hull, propeller, rudder and the wind propulsors.
The hull and rudder forces were determined by captive CFD computations. Based on this, a mathematical model for the hull was derived in 4 degrees of freedom (X,Y,K and N) which were a function of drift, heel and rotation. The open water diagram of the propeller was available to model the propeller thrust. The rudder forces were modelled by deriving the flow (including influence of the hull and propeller) and the lift and drag (accounting also for "lift carry over" to the hull).
At last, the wind forces of the Dynarigs were modelled by wind coefficients as function of the AWA. These wind coefficients were derived from the wind forces and moments measured in the zig-zag experiments.

The following simulations were done::

- Wind modelling identical to the experiments, where the AWS was considered to be constant and the yaw checking angles were equal to what was used in the experiment. Also identical initial rotation at the start of the manoeuvre was introduced. The purpose of these simulations was to see if the tests could be replicated.
- Secondly, the AWS was kept constant and corrected yaw angles were considered. Also the rotation at the start of the manoeuvre was set to zero.
- Thirdly, a more realistic wind modelling was applied. Here, the TWS and TWA were kept fixed (relative to the initial course).
- At last, simulations were conducted with no external wind (to represent a ship without WASP).

It can be seen that the first overshoot angle can be accurately reproduced with the simulations (within 1 degree). However, the second overshoot turns out to be less accurate compared to the experiments. The difference is respectively 6 and 9 degrees. A reason for this mismatch may be that the zig-zag manoeuvre is rather sensitive for the magnitude and direction of the wind forces. Over time, the error accumulates, which makes it harder to properly predict the second overshoot.

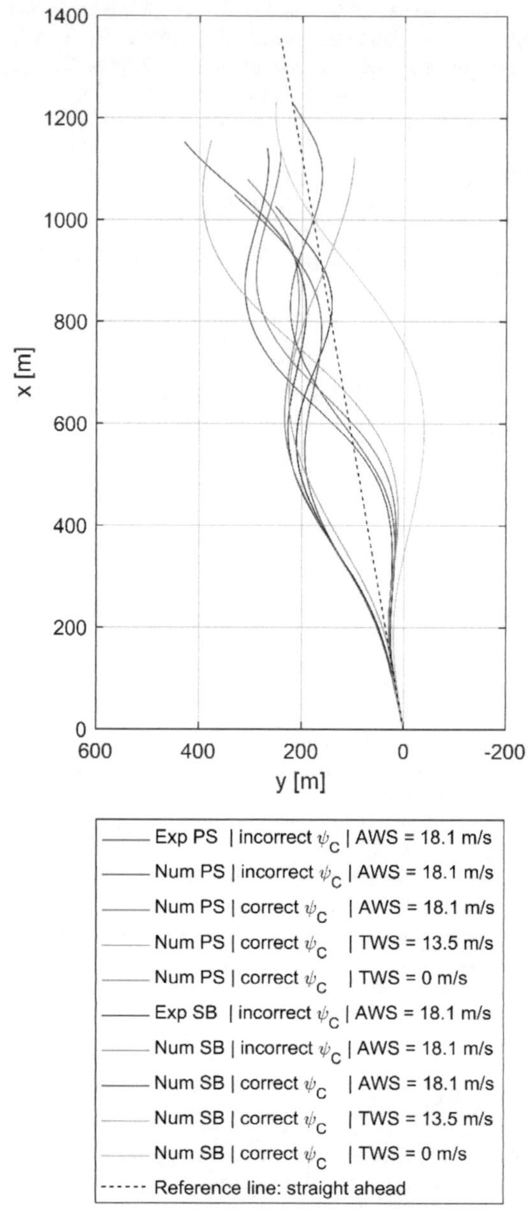

Figure 13: Bird eye view of experiments and simulation results

From the simulation results, it was found that the incorrect yaw checking angle had some influence on the overshoot angles. The overshoot angles differ between 0.4° and 2.1°.

When forcing constant true wind speed and angle, as opposed to the fixed apparent wind speed and angle, it is observed that on average the overshoot angles increase slightly. The initial turning ability decreases marginally.

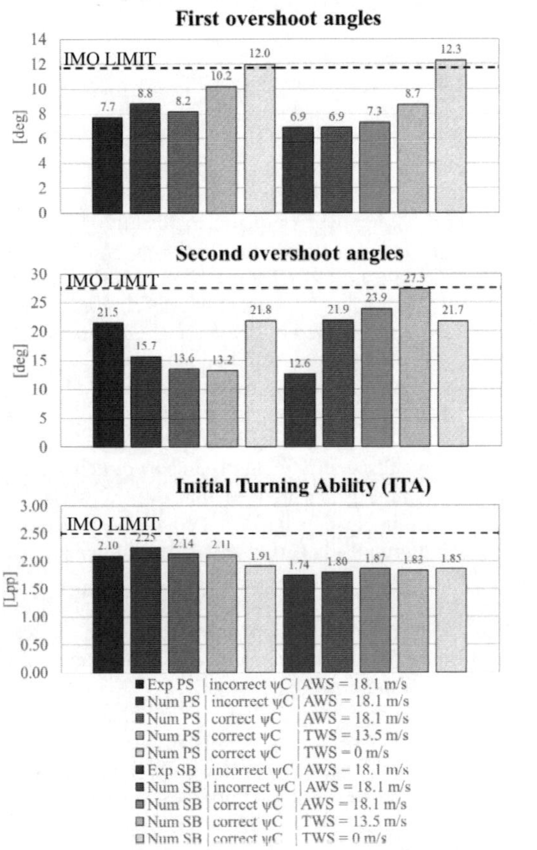

Figure 14: Manoeuvring characteristics derived from zig-zag 10°/10° manoeuvre and compared to IMO resolution MSC 137(76)

At last a comparison is made between simulations with a constant TWS and without wind. The first overshoot with wind results in lower first overshoot angles. For example, the zig-zag to starboard showed a first overshoot of 10.2° with constant TWS and 12.0° when no true wind is present. For the second overshoot, the zig-zag to port side is much improved in contrast to simulation with no wind. The second overshoot to starboard deteriorates with wind in comparison to no wind.

Observing the manoeuvre to port side, it is shown that initial turning ability deteriorates with $0.2 \cdot L_{pp}$. For the manoeuvre to starboard the initial turning ability is not much affected.

Overall, the manoeuvring characteristics, for the chosen wind conditions and Dynarigs does not generally improve or worsen. It depends on the particular characteristic that is looked at and the direction of the manoeuvre. However, when the worst case is selected from both port side and starboard, then the end result is inevitably worse.

7.4 ROLL DAMPING

Roll decay tests were done to determine the combined hydrodynamic and aerodynamic roll damping. Figure 15 shows the result for the vessel with Dynarigs expressed as percentage of critical damping, linearized for a roll amplitude of 2 deg. The left most bar shows the damping at zero ship speed with the Dynarig yards positioned perpendicular to the ship centreline, thus not providing any significant roll damping. An increase in damping is already seen in the next two bars when the Dynarig yards are set perpendicular to the ship centre line. When sailing without wind, damping increase substantially because of higher hydrodynamic damping and a bit of aerodynamic damping because of the ship's own speed. However, the damping increases further thereafter when switching on the wind fans (the right most bar). Although the exact magnitude is uncertain (because of very limited oscillations in the test), it is clear that the damping reaches values that are otherwise not obtainable for commercial vessels without special roll reducing devices.

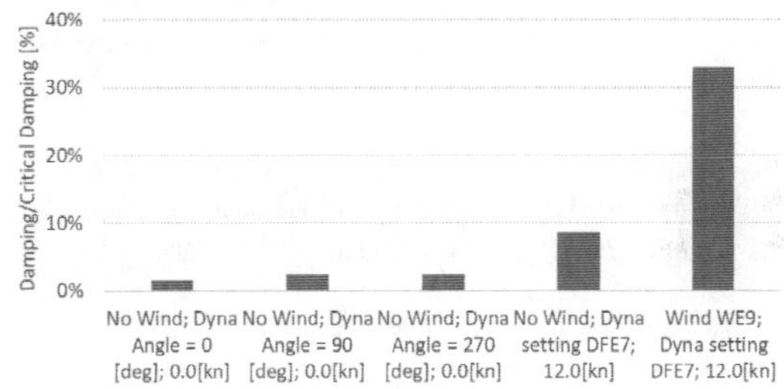

Figure 15: Roll damping as percentage of critical damping

Roll decay tests with Flettner rotors also show an increase in damping. However, the increase is modest in comparison to the hydrodynamic damping.

7.5 SEAKEEPING TESTS WITH FLETTNER ROTORS IN STERN QUARTERING SEAS

Test were conducted in irregular and regular stern quartering waves with the Flettner rotor. One test series focussed on regular waves with a wave amplitude of 2 m and a relative wave heading of -21 deg (waves incoming from the port side stern quarter), a series of wave periods and two different wind speeds. A single test was conducted at a wave heading of -25 deg.

In Figure 16 the test conditions are shown in relation to theoretical resonant roll and projected wave length along the hull being similar to 1.0 to 1.2 times the ship length. The latter is a measure of the yaw excitation. The test conditions are judged to be demanding in terms of the wave period and wave height. However, the roll and yaw excitation would have been bigger at a relative wave heading around 45 deg. However, such a heading would have resulted in shorter runs in the test set-up.

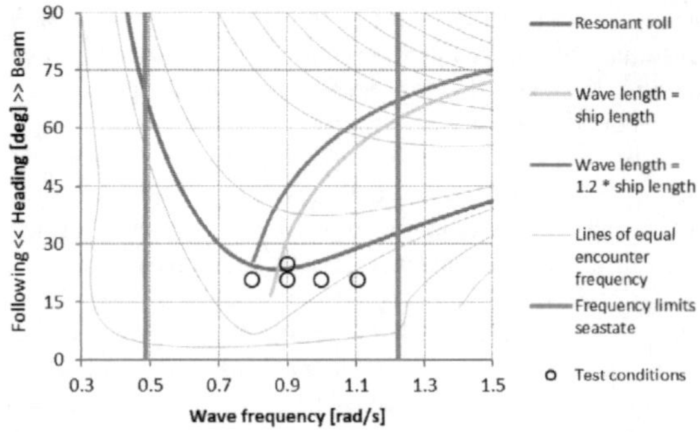

Figure 16: Wave conditions in regular seas

The purpose of these tests was to observe the changes mainly in course keeping as a function of the thrust of the wind propulsors. Therefore the tests were run for two different wind speeds, yielding two different levels of thrust.

Figure 17 shows the total Flettner thrust. It can be seen that the higher apparent wind speed resulted in a significantly higher thrust force (up to a wind assist ratio of about 50%).

Figure 18 shows the standard deviation of the yaw angle. It can be seen that the increased Flettner thrust (and decreased flow over the rudder) results in a much larger heading variation. Thus, the hypothesis that wind assistance may be a challenge in course keeping in stern quartering seas is confirmed. Further work should show whether it actually becomes problematic for operations, to see what is the response at other wave headings and how it could be mitigated using changed control or design.

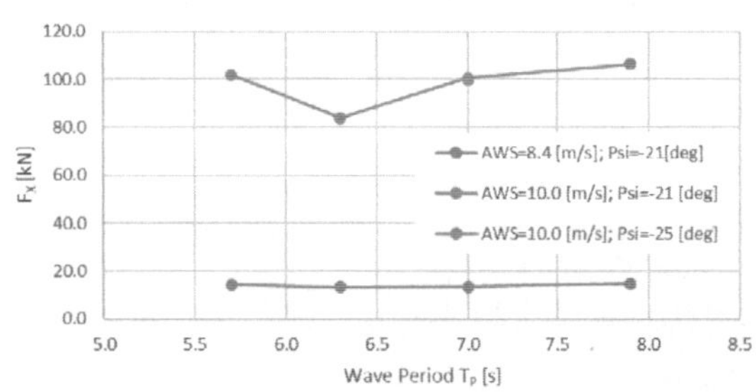

Figure 17: Total Flettner thrust as function of wave period

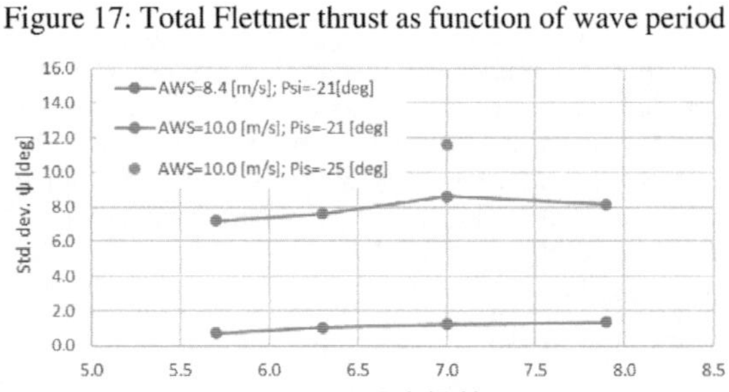

Figure 18: Standard deviation of yaw as function of wave period

8. CONCLUSIONS & RECOMMENDATIONS

The following conclusions are reached with regard to the test setup allowing to model aerodynamics and hydrodynamics at the same time:

- A reasonable wind field can be produced for close hauled conditions, although some twist is observed. The wind field is much more challenging in stern quartering conditions.
- The forces for both the Dynarigs and Flettner rotor fell short of what can be expected based on other reasonable data. The difference cannot fully be explained by scale effects. Although not confirmed, blockage and directional variations in the flow may be the reason for this. Preferably when using this set-up the forces are calibrated to known accurate references by slightly changing the wind speed or wind propulsors dimensions.
- In summary it is judged that the test set-up is usable in close hauled wind conditions when the wind propulsors forces are calibrated to a known reference.

The following first observations are made for the performance of ships with wind assistance in unsteady conditions:

- Manoeuvring with Dynarigs in closed hauled conditions showed to be significantly affected by the wind. The yaw checking and initial turning ability can either improve or deteriorate depending to which side the ship is manoeuvring. An rudder angle of 15° was considered to be the initial rudder angle, however the IMO does not clearly specify what rudder angle should be used when conducting zig-zag (and turning circle) manoeuvres with wind-assistance.
- The aerodynamic roll damping from Dynarigs is larger than the hydrodynamic roll damping from the hull, resulting in a significant reduction of roll motions. A smaller effect is seen for Flettner rotors.
- Course oscillations in stern quartering conditions with Flettner rotors can become much larger when using a large contribution of wind assistance.

The observations from this project are judged to be only a first scan. Further work should generalise the conclusions to other designs and conditions and to show how wind assisted ships can be best designed for unsteady conditions. The possible operational limitations as found from that research should then also be accounted for in performance predictions.

9. ACKNOWLEDGEMENTS

This research is partly funded by the Dutch Ministry of Economic Affairs".

MARIN wishes to thank Dykstra Naval Architects, EcoFlettner, Hochschule Emden-Leer and Norsepower for helping with the specifications and design of the wind propulsors on the scale model

MARIN wishes to thank the Peutz Group for conducting wind tunnel tests within the project.

10. REFERENCES

1. EGGERS, R., "Operational Performance of Wind Assisted Ships", 10th Symposium on High-Performance Marine Vehicles, 2016.
2. EGGERS, R., GAILLARDE, G., "Wind Assisted Ships Design Exploration and Operational Constraints", Power & Propulsion Alternatives for Ships, 2017
3. CE Delft, "Study On The Analysis Of Market Potentials And Market Barriers For Wind Propulsion Technologies For Ships" 2017
4. MARIN, "Design Variations for Wind Assisted Ships", Report 70033-10-SHIP, 2018
5. EGGERS, R., "Prediction Methods for team AkzoNobel in the Volvo Ocean Race and Wind Assisted Ship Propulsion for Merchant Vessels", 25th International HISWA Symposium on Yacht Design and Yacht Construction, 2018
6. SCHWARZ-RÖHR, B. et al., "Roadmap for Sail Transport: Engineering", NSR SAIL Project, 2015
7. BORDOGNA, G., KEUNING, J. A., HUIJSMANS, R. H. M., BELLOLI, M., "Wind-tunnel experiments on the aerodynamic interaction between two rigid sails used for wind-assisted propulsion," International Shipbuilding Progress, vol. 65, no. 1, pp. 93–125, Jun. 2018
8. INTERNATIONAL MARITIME ORGANIZATION, "Standards for Ship Manoeuvrability", MSC.137(76), 2002

11. AUTHORS BIOGRAPHY

Rogier Eggers holds the position of Senior Project Manager and Knowledge co-ordinator of Ships Manoeuvring at the Maritime Research Institute Netherlands. His main business is to conduct model tests and simulations and to help clients regarding manoeuvring and seakeeping of ships. Since 2011 his work also includes research regarding Wind Assisted Ship Propulsion.

Anton Kisjes holds the position of Project Manager at the Maritime Research Institute Netherlands. In 2018 he graduated with his thesis about wind-assistance for merchant vessels. At MARIN his work consist of conducted manoeuvring experiments and simulations to help clients with improving and understand the manoeuvrability of ships.

Wind Propulsion, 15th -16th October 2019, London, UK

ZERO EMISSION SAILING SHIP – CONCEPTUAL DESIGN –

K. Ouchi, Ouchi Ocean Consultant Inc., Japan and **T. Omiya,** Mitsui O.S.K. Lines, Ltd., Japan

SUMMARY

When a sailing ship which has large rigid wing sails such as the Wind Challenger Sail runs in a sufficiently windy sea, the thrust by sails is utilized to not only drive the ship at the proper speed but also to rotate an underwater turbine at significant speed and torque. The turbine generates electricity which is used for the electrolysis of water to generate hydrogen. The hydrogen is stored using toluene in the form of methylcyclohexane (MCH), which is in liquid form under normal temperature and pressure. MCH is stored in the ship's tank as hydrogen fuel. In the case of weak winds when the sails cannot generate sufficient thrust, using the hydrogen generated by the dehydrogenation device, the fuel cell works and supplies electricity to the electric motor propeller for the ship's propulsion. Thus, the ship can run at a constant speed regardless of wind speed and direction.

1. INTRODUCTION

In order to realize the sustainable society, CO_2 emission from the activities of mankind needs to be reduced drastically.
Following the 2015 Paris Agreement, at the April 2018 IMO International Marine Transportation Meeting, it was resolved that all maritime voyages to emit zero CO_2 within this century, and to halve CO_2 emission by the year 2050. This resolution will strongly accelerate the change of ship's propulsion energy source from the fossil oil to zero emission energy source. The innovation of zero emission ship propulsion system needs to be researched and realized urgently.

2. CANDIDATES OF ZERO EMISSION SHIP

Table 1 shows features of conventional ships and candidates for zero emission ships. It also shows their energy source, power generation device, propulsor device and remarks.
As for nuclear ships, a practical technique has already been established and vastly applied in naval ships and submarines. It is not so difficult technically to build nuclear cargo ships, however, it is difficult to introduce nuclear merchant ships in the international shipping industry because of the negative opinions of the general public regarding their safety and possibility of nuclear disaster. Regarding the CO_2 recovery ships, the main part of the ship propulsion system is the same as that of conventional ships, however, the problem lies not only with the size of the additional CO_2 recovery device, but also with the cost of transporting CO_2 and building the CO_2 storage site. Furthermore, the general public acceptance of CCS (Carbon Capture and Storage) is also problematic. As for battery-operated ships, even in the case of the most advanced lithium battery, the battery weight is too heavy, and it is not suitable as the propulsion system for high-powered long-distance ships. Furthermore, the battery needs to be charged by electricity originating from a renewable energy, so called CO_2 free electricity, whose cost is very high. Both fuel-cell ships and hydrogen engine ships are already in use in the maritime community. However, their technology is not yet popular because of the high cost of CO_2 free hydrogen.
Among the above candidate ships, we would like to recommend ships powered by wind as the main basis for the zero emission ships. They have been utilized and have been accustomed to the ocean winds since the ancient times. Sailing ships enjoyed their golden era in the 19th century, however, wind speed and direction on the seas are not constant, so that it has been impossible to keep ships on punctual schedules. Since the industrial revolution, especially after the invention of the steam turbine and diesel engine, steam ships and diesel ships, which run without wind, have replaced sailing ships.
In this paper we would like to propose a "Zero Emission Sailing Ship" (ZESS) which runs by wing sails, and generates hydrogen by a water turbine generator and water electrolyser in strong wind areas. In weak wind

Table 1 Candidates of Zero Emission Ship

Kind of Ship	Energy Source	Power Generation	Propulsor	Remarks
Conventional	Oil , Gas	Diesel Engine	Propeller	Much CO2 Emission
Nucliar	Uranium	Steam Turbine	Propeller	Uranium Supply Limit, No Public Acceptance,
Wind Powered	Ocean Wind	Wing Sail	Wing Sail	Few Thrust in Weak Wind
Hydrogen engine	CO2 Free Hydrogen	Hydrogen Gas Engine	Propeller	CO2 Free Hydrogen Supply, Emission of NOx
Fuel Cell	CO2 Free Hydrogen	Fuel Cell	E. Motor & Propeller	CO2 Free Hydrogen Supply
Battery	CO2 Free Electricity	Lithium Battery	E. Motor & Propeller	Small Capacity for Long Term Voyage
Bio Fuel	Biofuel	Diesel Engine	Propeller	High Cost Fuel, Pressure on Cultivated Land for Food
CO2 Recovery	Oil, Gas	Diesel Engine	Propeller	Needs of CO2 Reccovery Plant

areas, it will run using the hydrogen energy which has been generated and stowed in strong wind areas. In this way all the energy needed for the ship is captured from the wing sail, i.e. the ship consumes no oil and emits no CO_2.

3. WIND CHALLENGER PROJECT

We are now undertaking the Wind Challenger Project to develop an extraordinarily large rigid wing sail (height: 60m, breadth: 15m, thickness: 3m, made of GFRP). This rigid sail can be rotated for the wind direction and be retracted telescopically in case of sail reefing in too strong winds [2].

Fig.1 shows an 84kDW bulk carrier fitted with one Wind Challenger Sail on the forecastle of the ship. The fuel saving ratio of the ship is estimated at an average of 4 - 7%, and in the case of four fitted Wind Challenger Sails will save the four times fuel savings at an average of 14 - 28%. Fig.2 shows the reef mode of the Wind Challenger Sail retracted, which is used in rough sea condition or in port. The first Wind Challenger Ship will be delivered to the owner (Mitsui O.S.K. Lines) from the shipyard (Oshima Shipbuilding) in the 2nd half of 2022FY.

Fig.1 84kDW Bulk Carrier fitted with the Wind Challenger Sail (Full Sail)

Fig.2 84kDW Bulk Carrier fitted with the Wind Challenger Sail (Reef Sail)

4. ZERO EMISSION SAILING SHIP

To achieve zero emission from the sailing ship, an acquisition of a large amount of wind energy is important. The size and performance of the Wind Challenger Sail should not only be enough to run the ship at the cruising speed, but also to generate and store a proper amount of hydrogen. The volume of hydrogen gas is large; therefore, it is necessary to change hydrogen gas into another form such as liquid hydrogen, Methyl-Cyclo-Hexane (MCH), ammonia, etc. In this project, the MCH/Toluene system may be chosen for safety and economy. A water turbine/generator which absorbs the kinetic energy of relative water flow is installed, generates electricity, and activates the water electrolysis device [1]. The water turbine/generator is utilized as propeller/electric motor when the water turbine/generator rotates in normal or high pitch.

Fig.3 shows a summary for the Concept and Energy Flow of the Zero Emission Sailing Ship. The ship has two methods for propulsive operation:

1) In the case of strong winds, indicated by the red line in Fig.3, the excessive energy, i.e. that produced exceeding the energy needed for ship's cruising speed, generates electricity, which then activates the electrolysis device and produces hydrogen gas. The hydrogeneration device enables hydrogen gas to be attached to toluene and makes MCH act as the hydrogen carrier and stores it in the MCH tank
2) In the case of a weak wind, indicated by the blue-dotted line in Fig.3, the energy deficit in the energy needed for the ship's cruising speed is supplied by the fuel cell via the electric motor driven propellers. The fuel cell is driven by the hydrogen gas which is separated from the MCH stowed in the MCH tank

5. PARTICULARS AND ARRANGEMENT

Fig.4 shows the main particulars and rough arrangement for a 74kDW Bulk Carrier type Zero Emission Sailing Ship. The wheel house is located in the fore part of the ship in order to keep forward visibility and the accommodation and engine room is near by the wheel house. The hull form is so called a Buttock Flow type with large skeg in the ship's centre line. The system avoids the speed reduction by the hull wake and keeps course stability. Two pod type turbine/propellers with generator/ motors are installed at the end of the hull to collect or to give kinetic energy.

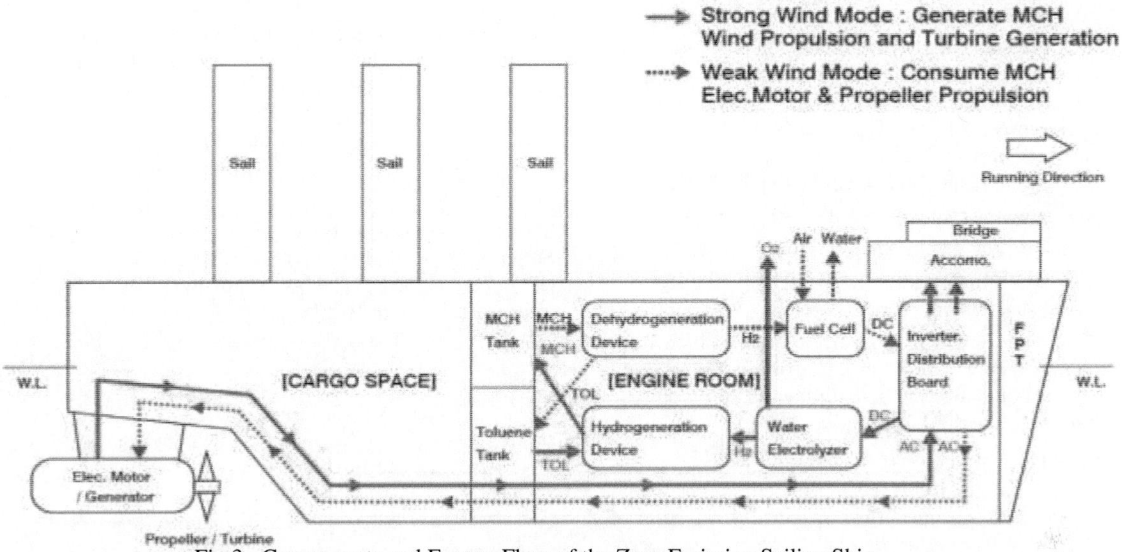

Fig.3 Components and Energy Flow of the Zero Emission Sailing Ship

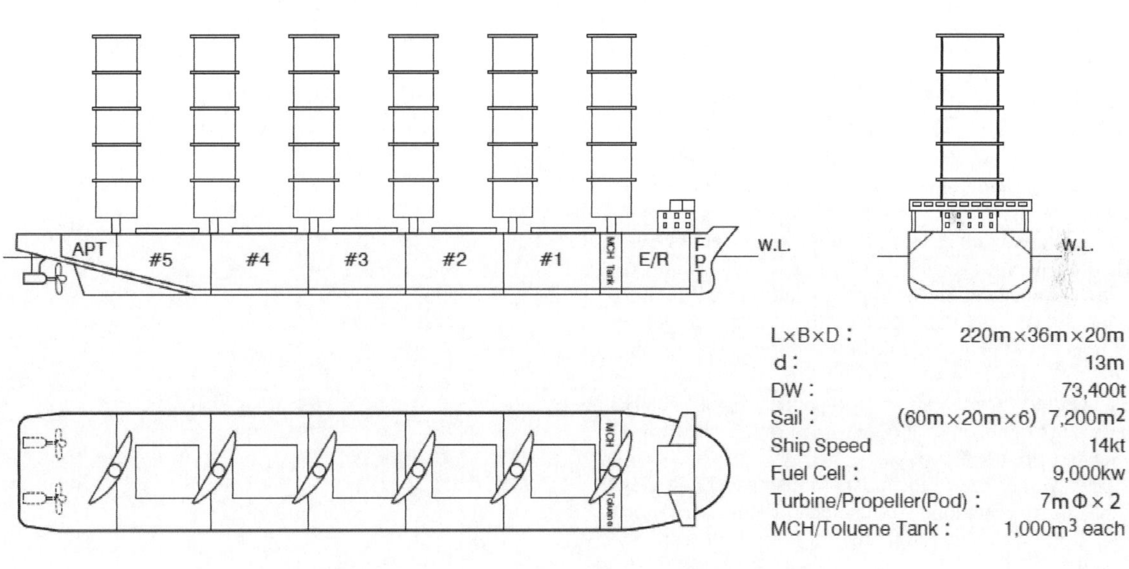

L×B×D :	220m×36m×20m
d :	13m
DW :	73,400t
Sail :	(60m×20m×6) 7,200m2
Ship Speed	14kt
Fuel Cell :	9,000kw
Turbine/Propeller(Pod) :	7mΦ×2
MCH/Toluene Tank :	1,000m3 each

Fig.4 Rough General Arrangement and Main Particulars

6. SUBJECT FOR FUTURE DISCUSSION

The following is the agenda for future discussions towards the realization of the Zero Emission Sailing Ship (ZESS):

・Selection of a hydrogen carrier

Chemical hydride (Toluene/Methyl-Cyclo-Hexane) is used here, but alternatives such as liquid hydrogen, pressurized hydrogen, ammonia, and hydrogen-absorbing-alloy can also be considered. A discussion to compare their safety, cost, and ease of handling is required

・Weight reduction of Wind Challenger Sails

To enable increasing the area and height of Wild Challenger Sails, the adoption of light and strong materials such as CFRP for moving spars inside of the wing sail is needed instead of steel spars.

- Common use of Generator/Motor and Turbine/Propeller

  The ship uses a hydrogen production mode in strong wind areas, and hydrogen consumption mode in weak wind areas. Therefore, a turbine/propeller device is used as an energy collector in strong winds, and an energy emitter in weak winds. Research into the optimum blade shape will be required.
- Customizing the fuel-cell specification

  Development of the fuel-cell to meet the power demand of the two-mode operation will be needed.
- Integral usage of heat energy

  Heat energy which is absorbed in or emitted from components such as fuel-cells, water electrolyser, hydrogeneration, and de-hydrogeneration should be researched and used properly.
- Voyage plans and selection of voyage routes

  Development of voyage programs is necessary, taking into account the punctuality of voyage schedules, the weather forecast, ship cruising speed, and safety.

7. CONCLUSION

We propose the Zero Emission Sailing Ship (ZESS) as a zero CO_2 emission ship abiding by the resolution of the 2018 IMO committee. The total energy for the ZESS is supplied by the ocean wind and ZESS has two modes of operation: the energy collection mode in strong winds; and energy consumption mode in weak winds. The water turbine (commonly used with propellers), generator (commonly used with electric motors), water electrolyser, and hydrogeneration device will be used to collect energy. The de-hydrogeneration device, fuel-cell, electric motor (commonly used with generators), and propeller (commonly used with water turbine), will be used to emit energy.

The technology for the above components has basically already been established and there will be no need for a new technology breakthrough.

The ZESS will be one of the most promising technologies for the next generation of merchant ships, and an acceleration of its research and development can be expected.

8. REFERENCES

[1] Ouchi, K., and Henzie, J. 2017. Hydrogen Generation Sailing Ship. The proceedings of OCEANS2017 Aberdeen Scotland, IEEE & Marine Technology Society.

[2] Ouchi, K., Uzawa, K., Kanai, A. and Katori, M. 2013. "Wind Challenger" the Next Generation Hybrid Sailing Vessel. Proceedings of The Third International Symposium on Marine Propulsors (smp'13) Launceston, Tasmania, Australia.

9. AUTHORS BIOGRAPHY

Dr Kazuyuki Ouchi: CEO, Ouchi Ocean Consultant Inc. He is responsible for R&D Adviser for Mitsui O.S.K Lines, especially on Wind Challenger Project, Hydrogen Generation Sailing Ship and Zero Emission Sailing Ship. His previous experience includes Professor of Graduate school of Frontier Sciences, The University of Tokyo

Mr. Tomoki Omiya: Smart Ship Strategy Team, Smart Shipping Division, Mitsui O.S.K. Lines, Ltd.
He received B.S. and M.S. degrees of mechano-informatics from the University of Tokyo.

TRANSITION PATHWAYS TO VERY LOW EMISSIONS SHIPPING: THE MATISSE-SHIP MODEL

J. Köhler, Fraunhofer ISI, Germany

SUMMARY

The MATISSE-SHIP simulation model of transitions has been developed to produce scenarios of technical change. An agent-based approach represents investment decisions of shipowners in cruise/Ro-Pax, container, dry and wet bulk and service vessels and technologies including wind assistance, wind hybrid and alternative fuel motors. Decision parameters include: capital and operational costs, GHG and SOx/NOx intensity and requirements for new bunker infrastructure. The attitudes of shipowners towards operational changes such as weather routing and autonomous ships are represented.

The policy scenario extends the IMO NECAs, SECAs and 2020 Global sulphur cap to include effective policies for GHG emissions reductions implemented in a 2025-2040 time frame. Wind becomes more competitive than the low flashpoint fuels, which have higher fuel costs and higher GHG emissions, while power to liquid fuels from renewables may have a market in trades requiring very high power outputs if costs come down and bunker infrastructure is developed.

NOMENCLATURE

ABM	agent-based model
GHG	greenhouse gas
IMO	International Maritime Organisation
LfL	Low Flashpoint Liquid fuel
LNGDF	Liquid natural Gas Dual Fuel motor
NECA	NOx control area
PtL	power-to-liquid synthetic fuels
SCR	Selective Catalytic Reduction NOx removal technology
SECA	SOx control area

1. INTRODUCTION

Shipping is a major contributor to global greenhouse gas (GHG) emissions [1]. A further major environmental problem is the emission of SOx, NOx and soot from ships, which has led to the implementation of emissions control areas for NOx and SOx, the so-called IMO NECAs and SECAs [2] Furthermore, the IMO has agreed a global SOx cap to be enforced from 2020, with a limit of 0.5% Sulphur emissions from all ships over 100 Tonnes (GRT). ECAs are now being enforced on the West and eastern coasts of the USA, in the North Sea and Baltic, in China and Japan.

there is a wide range of alternative technologies available for ship propulsion and power systems that can reduce emissions. However, there is a great deal of uncertainty about the direction of change that commercial shipping can or should take. While LNG and other LfL fuels do offer very large reductions in NOx, SOx and particulates, they impose increased costs compared to diesel engines fuelled by HFO or MGO. Furthermore, there are very few scenarios of future shipping that illustrate how reductions of 80% or more in GHGs can be achieved.

The MATISSE-SHIP simulation model of transitions has been developed and calibrated to produce scenarios of transitions in shipping. It is based on the MATISSE transitions model [3,4], which uses the Multi-level Perspective (MLP) theoretical framework of niches, regime and landscape in an agent based model (ABM) simulation to generate potential transition pathways. This explicitly combines economic considerations with institutional and attitudinal factors to determine the choice of system. The MATISSE model was developed with a general structure, which can be interpreted for different contexts.

The paper describes the structure of the MATISSE model and its adaptation and interpretation for shipping. The calibration of the model is explained and some initial results for transition pathways to alternative propulsion technologies presented. These results suggest scenarios and conditions under which a transition to wind technologies or to alternative fuels might develop.

2. The MATISSE-SHIP model

The conceptual structure of the model is described in detail in [5] and [3] is a detailed description of the model. The MATISSE model structure is illustrated in figure 1.

The idea is to apply the recent theoretical developments in innovation studies to allow for the complex co-evolutionary processes in large scale socio-technical-environmental systems [5].

The regime and niches (see figure 1) are production networks - decision making agents who provide propulsion technologies and ships (we shall simplify this to call them shipyards). Other agents represent the demand side, ordering new ships and potentially rebuilds i.e. shipping companies or long-term charterers. The macro-level Landscape represents changes in society generally, which may exert pressure on the regime and/or support the development of niches.

For the case of shipping, the regime is considered to be propulsion by *diesel* motors. The alternatives considered for niches are:

Liquid Natural Gas including dual and triple fuel engines (LNGDF);
Low flashpoint Liquid fuels e.g. Methanol, ammonia, dimethyl ether (LfL);
hydrogen;
wind;
wind assistance (windassist).
biofuels;
synthetic fuels (so-called power-to-liquid PtL).

Operational measures are discussed under practices below.

The model uses the concept of practices as the metric through which agents position themselves in society and over which behaviour is defined. Practices are each represented as values along axes, constituting a multi-dimensional *practices space*.
Agents - the technologies supplied and shipping companies are differentiated by their positions in the multi-dimensional practices space.
Practices included in the MATISSE-SHIP model are:

- GHG intensity - CO2 emissions/tonne km or passenger km;
- fuel cost Euro/tonne;
- operational speed (adoption of slow steaming);
- local air emissions /tonne km or passenger km (NOx, SOx, particles);
- capital cost/MW;
- perceived technological and operational change (reduced operational speed and weather optimised routing, even if this involves extended transit times compared to diesel propulsion at the operational speeds of 2012;
- requirement for new bunker infrastructure.

Both shipyards and shipping companies can change their technologies, shipyards according to their success in selling ships and shipping companies according to their requirements, defined as the values of the practices they wish to achieve. The regime - diesels - develops its technology to maximise the number of new ships sold over all shipping companies. The niche technologies try to meet the requirements of shipping companies whose requirements are close to what their technologies can provide.

In each year, the shipyards see whether their sales are increasing and if they are not, change their technologies to try and obtain more orders. They are restricted in how fast and in which direction they can change. The regime can only change slowly, while the niches can develop more quickly, but only in a limited range of directions (e.g. to reduce greenhouse gas emissions for low carbon technologies such as power to liquids or wind, or lower NOx/SOx for LNG). See [3,4] for details.

The shipping companies decide whether they will order a new ship and then if they do, choose the technology that is currently closest to their requirements measures as values of the practices.

The results of the model are then the shares of the competing shipyards of orders for new ships from the shipping companies. In this way, the model simulates technology diffusion through time, running from 2015-2050.

3. CALIBRATION OF THE MODEL AND TRANSITION PATHWAYS

3.1 INITIAL CALIBRATION

Each agent is individually parameterised (in the MATISSE-SHIP model over 1000 shipping company agents) as well as the niches and regime. Because shipping technologies and operations are global, is necessary that the model reflects the structure of the global fleet and the various general types of ships in operation. In order to enable a relatively small and flexible model, which can address whole system changes, the model is a simplified an aggregated representation of the world fleet. The model is calibrated using the fleet data from [1]. The initial values of the practices for the technologies are shown in table 1 and the initial practices for the shipping company markets are shown in table 2. Practice dimensions are normalised to scale of 1 to 100, giving equal weighting of all practices distances in the practices space.

3.2 SCENARIOS OF TRANSITION PATHWAYS

Scenarios of transition pathways are generated as the development of the technology shares through time. The paths of technology shares are determined by the choices of the shipping companies for new investments. Initially (in 2012-2015) almost all shipping companies in all trades used diesel engines. The only significant exception, apart from military vessels is the LNG (and a few other liquid bulk carriers), where a large part of the fleet uses dual fuel LNG/diesel motors. There are a small number of LNG ships in operation, in particular for cruise ships operating in NECA and SECA areas [2].

The pathways of change are generated by changing requirements for shipping operations. Currently, the strongest drivers of change are the IMO NECA and SECA areas and the prospect of the global sulphur cap in 2020 [2]. In the longer term, it is possible that climate policy will deliver stronger requirements for reducing GHG emissions. A further possible change is that the relative prices of fossil fuels and low carbon alternative fuels will change. Two preliminary scenarios illustrate the possible dynamics that may arise from such developments.

Figure 2 illustrates a scenario that results from a requirement for the shipping companies to drastically reduce NOx and SOx emissions i.e. a change in the NOx SOx practice from 100 towards a value of 5. After 2020, shipping companies begin to move away from conventional fossil fuels, because of the costs of scrubbers and SCR emissions reduction equipment. In the current parameterisation, wind propulsion is successful because it has a much lower fuel requirement and therefore a considerable cost advantage over the other alternative technologies. Although LfL is the most successful initial technology, the somewhat higher cost of motors and cryogenic tanks and fuel systems proves to be a decisive disadvantage. Some operators adopt wind propulsion, assuming that quite a few operators are willing to change their operating practices to reduce ship sizes in bulk trades. The major advantage of wind propulsion is that it offers considerable reductions in operating costs on those routes where the prevailing distribution of wind patterns enables a regular service. This trend could be reinforced by further automation of ships, such that the increased length of time at sea for the lower operating speed is not a disadvantage in terms of finding crews or wage costs per voyage.
However, LfL fuels are also adopted because they offer the necessary emissions reductions and the infrastructure can be expanded in a similar way to LNG, extending already existing industrial infrastructure, but at a lower cost than LNG. Fuel prices are also assumed to change in parallel to LNG. Because there is no strong climate policy, LfL fuels increase in competitiveness as they are adopted on a larger scale, bring down the fuel price and capital expenditures. Eventually, they out compete wind, as more shipping companies wish to keep their higher operating speeds and voyage times.

Figure 3 shows what might happen if, in addition to the IMO NECAs, SECAs and 2020 Global sulphur cap, effective policies for GHG emissions reductions are implemented in a 2025-2040 time frame. In addition to the change in the NOx SOx practice dimension, the GHG practice for shipping companies also changes away from the initial value of 100 towards a much decreased value of 10. In the long run, wind then becomes more competitive than the LfLs, which have higher fuel costs and higher GHG emissions. However, in the longer run, GHG neutral PtL fuels reduce their costs and start to become competitive. PtL fuels also offer very low GHG emissions, but do not have the power limitations of wind rigs. The main disadvantages of PtL fuels are currently that they are expensive fuels and that a large-scale supply chain has not yet been established. Therefore, it will take some time for them to become competitive, even with effective supporting environmental policies.

4. DISCUSSION AND CONCLUSIONS

These preliminary results show that a model that represents the niche-regime-landscape structure of the multi-level perspective on transitions can be used to simulate sustainability transitions in shipping. An important point is that the model dynamics are not just determined by the costs of the technological options and the emissions regulations, although these are major determinants. The requirements for changes in ship operations is also an important factor. The results that LfL and wind are the first two options chosen are also surprising, given that in the industry, LNG and scrubbers are regarded as the two main options for meeting the impending IMO global Sulphur cap [2]. The increasing cost of conventional diesel propulsion with exhaust treatment systems reduces the competitiveness of diesel engines, leading to a gradual move away from the diesel propulsion regime.
The number of LNG vessels in operation and on order is still a very small proportion of the global fleet. A major (perceived) issue with LNG is the availability of bunkering infrastructure and for this reason; the bunker practice value for LNG is set at 60, a significant departure from the conventional HFO/MGO bunker ships and infrastructure. However, LNG bunker vessels are now being built and many shipping operations with LNG can be covered either by mobile 'truck-to-ship' bunkering for small vessels or a few bunker vessels in the major international ports [2]. For other fuels such as methanol, where an infrastructure for general industrial use is already available, a specialist bunker infrastructure can expanded in a similar way, such that this is not a disadvantage for LfL fuels compared to LNG. The system cost advantage of LfL compared to LNG, where the fuel does not require such extreme low temperatures and specialist cryogenic systems is then, in this parameterisation, decisive.
The rapid uptake of wind extends the results of [7, 8]. It should be noted that the parameterisation of the wind practices does include the drastic changes in operating speed and operations management. However, the need for completely new ship designs for full wind operation and the necessary development period, combined with the need to gain experience with commercial operations with wind ships does suggest that these results are optimistic in terms of the speed of uptake. The results do suggest, however, that the underlying benefit of wind systems: that fuel costs are drastically reduced in addition to GHG reductions, can lead to a transition to sustainability in international shipping.
It is emphasised that these are preliminary results. Several aspects of the model can be enhanced. The market structure of shipping operations is highly simplified. The model does provide a structure in which the different trades and their requirements can be examined in more detail, both in terms of the requirements of different trades for the shipping operations and in terms of the modelling of innovation in both the regime and niches. The results do not illustrate the use of combinations of technologies (e.g. LNG and wind) in detail, although this is in principle possible in

the model through the amalgamation of niches. Also, battery propulsion and shore based electricity supplies for auxiliary systems have not been considered.
Finally, these results show that the range of potential pathways into the future is more complex than the usual results generated by techno-economic models of emissions mitigation technologies [6]. Currently (2019) it is not clear which technologies might come to be prevalent in a low emissions shipping system. The model generates a more realistic picture of the uncertainty facing the industry than conventional techno-economic models. At the same time, it is acknowledged that the model is highly simplified. The results should therefore be regarded as indicative scenarios and not as detailed projections for the future.

5. REFERENCES

[1] IMO (2014): Third IMO GHG Study 2014. MEPC 67/INF.3. Hg. v. IMO. London. Online verfügbar unter http://www.imo.org/en/OurWork/Environment/Pollution Prevention/AirPollution/Pages/Greenhouse-Gas-Studies-2014.aspx.

[2] Köhler, J.; Kirsch, D.; Timmerberg, S. (2018): Teilstudie „Studie über die Marktreife von Erdgasmotoren in der Binnen- und Seeschifffahrt". Wissenschaftliche Beratung des BMVI. Hg. v. BMVi (Bundesminiserium für Verkehr und digitale Infrastrukture: Fraunhofer Institute. Karlsruhe. Online verfügbar unter https://www.bmvi.de/SharedDocs/DE/Artikel/G/MKS/studie-marktreife-erdgasmotore-schifffahrt.html?nn=214206.

[3] Bergman, Noam; Haxeltine, Alex; Whitmarsh, Lorraine; Köhler, J.; Schilperoord, Michel; Rotmans, J. (2008): Modelling Socio-Technical Transition Patterns and Pathways. In: *Journal of Artificial Societies and Social Simulation* 11 (3:7). Online verfügbar unter http://jasss.soc.surrey.ac.uk/11/3/7.html.

[4] Köhler, Jonathan; Whitmarsh, Lorraine; Nykvist, Björn; Schilperoord, Michel; Bergman, Noam; Haxeltine, Alex (2009): A transitions model for sustainable mobility. In: *Ecological Economics* 68 (12), S. 2985–2995. DOI: 10.1016/j.ecolecon.2009.06.027.

[5] Haxeltine, A.; Whitmarsh, L.; Bergman, N.; Rotmans, J.; Schilperoord M.; Köhler J. (2008): A Conceptual Framework for transition modelling. In: *Int. J. Innovation and Sustainable Development* 3 (1-2), S. 156–161.

[6] Köhler, J. (2019): Advances in modelling sustainable innovation: from technology bias to system theories and behavioural dynamics. In: Boons F. McMeekin A. (Hg.): Handbook of Sustainable Innovation: Routledge.

[7] Nelissen, D.; Traut, M.; Köhler, J.; Mao, W.; Faber, J.; Ahdour, S. (2016): Study on the analysis of market potentials and market barriers for wind propulsion technologies for ships. Hg. v. CE Delft. Online verfügbar unter http://www.cedelft.eu/publicatie/study_on_the_analysis_of_market_potentials_and_market_.

[8] Köhler, J.; Nelissen, Dagmar; Traut, Michael (2017): Fighting the windbreak. In: *The Naval Architect* (May), S. 26–32.

6. AUTHORS BIOGRAPHY

Jonathan Köhler holds the current position of senior scientist at. His previous experience includes the position of Constructure, Royal Corps of Naval Architects and manager, Integrated Assessment Model Development for the Tyndall Centre.

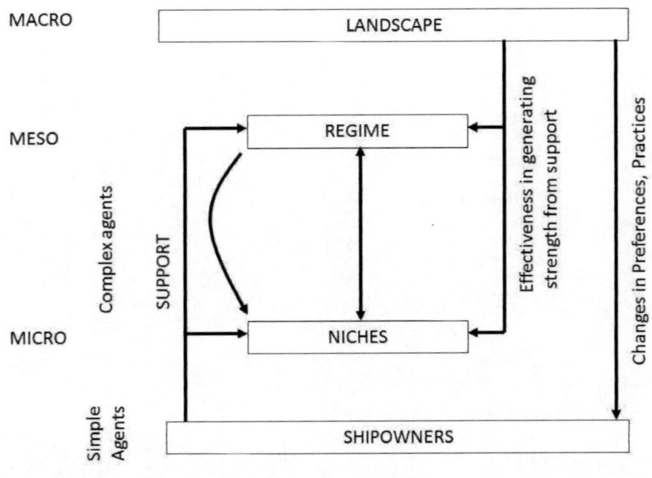

Figure 1: Structure of the MATISSE model.

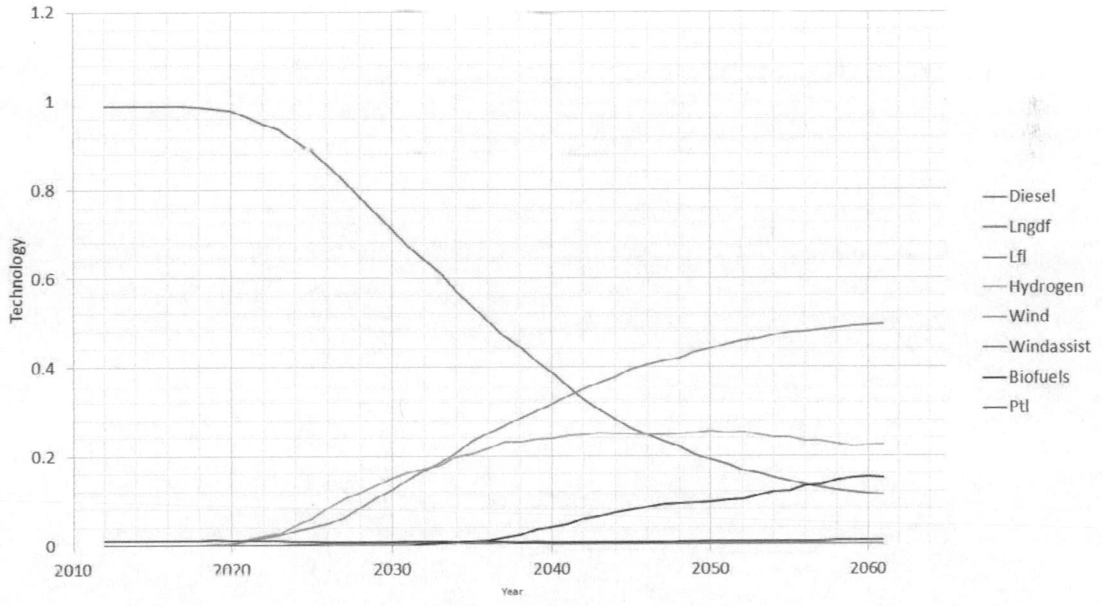

Figure 2 A 'NOx SOx' transition pathway to LfL fuels

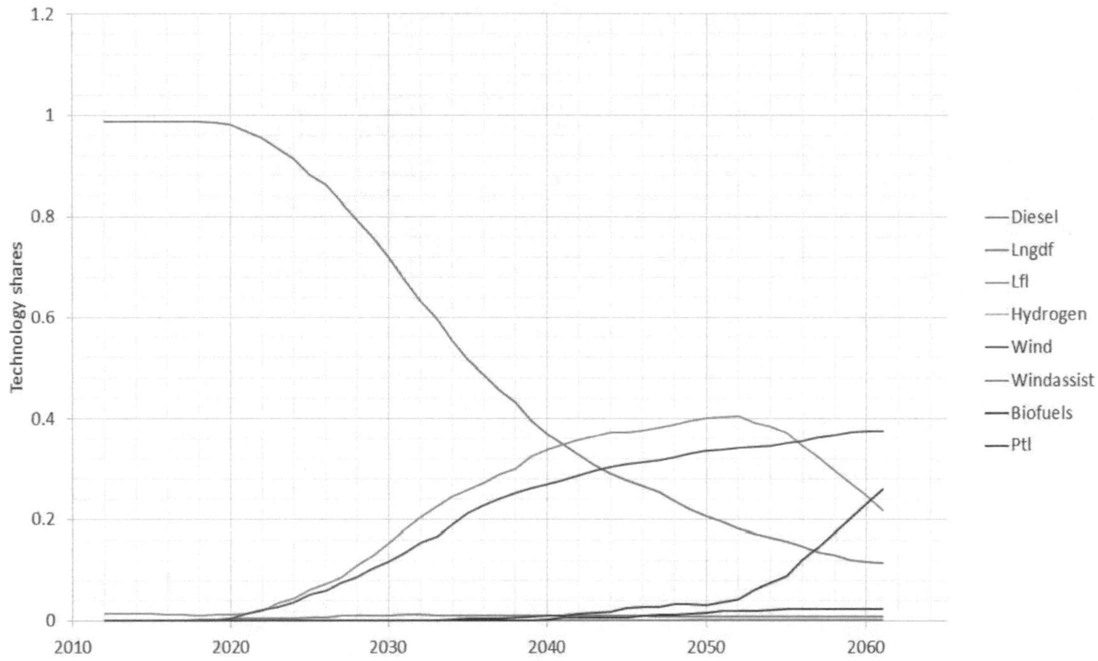

Figure 3 A transition pathway to wind propulsion

Table 1 Initial practices of technologies

	GHG Emission CO_2Equiv/ MW	Fuel Cost $/Tonne HFO	Opspeed Av. Kts	NOx SOx emissions[1] % of diesel	CAPEX $/MW	Operational and technology change	Bunkers - Compatible with current HFO/MGO
Diesel HFO+ MGO	90 [10,100] (0)	40 [0,100] (0.5)	50=25kts 10000TEU 15 Bulker [0,100] (0)	100 [0,100] (0)	40 [30,60] (0)	100 [50,100] (0)	100 [90,100] (0)
LNG (including dual fuel)	80 [70,100] (0)	38 slightly lower than ISO 380 [0,100] (0.5)	50 as diesel [0,100] (0)	10 (pure) LNG has no SO2 and almost no NOx [0,100] (0)	80 mature tech, LNG tanks very expensive [40,60] (0)	90[2] [0,50] (0)	50 mature but new cryo bunker infrastructure [60,80] (0)
Low Flashpoint Fuels e.g. methanol (including dual/triple fuel)	80 [70,100] (0) LNG feedstock	38 slightly lower than ISO 380 [0,100] (0.5)	50 as diesel [0,100] (0)	10 Methanol has no SO2 and almost no NOx [0,100] (0)	50 Mainly single-use zones [30,50] (-0.5)	90[3] [0,50] (0)	90 mod HFO bunker infra-structure [4] [90,100] (+0.1)
Hydrogen (Stored as H2 on board)	10 renewables;stored on board [5,100] (0)	80 [0,100] (0)	50 lower energy density than Diesel HFO [0,100] (0)	10 renewables [5,100] (0)	100 higher than LNG [0,100] (0.1)	100 [20,100] (0)	50 industry infrastructure mature[5] [50,90] +0.1)
Wind	5 [1,10] (0)	10 (for emergency/aux [5,40] (0)	10=5kts [5,20] (0.1)	10 (Aux/ emergency) [5,10] (0)	30 Mainly single-use zones [5,50] (-0.5)	10 v.slow and variable [5,10] (0)	100 aux only [0,100] (0)
Wind Assistance 10% energy saving of HFO	80 (90%) of ICE [50] (-0.5)	35 (half local, half long distance) [0,90] (-0.5)	45 optimally a bit slower [0,50] (0)	90 [40,100] (-0.5)	45 extra installation, smaller diesel [30,50] (-0.5)	90 in demonstration [5,100] (0)	100 [50,100] (0)
Biofuel	70 [0,100] (-1.0)	80 [50,100] (0)	50 as diesel [0,100] (0)	10 [5,100] (0)	40 drop-in [30,60] (0)	100 [50,100] (0)	80 drop-in, limited supply chain [90,100] (0)
PtL/PtG Power to Liquid/Gas I renewables	30 [5,50] (-0.1)	80 [6] [50,100] (0)	50 as diesel [0,100] (0)	10 [5,100] (0)	40 drop-in [30,60] (0)	100 [50,100] (0)	80 drop-in, limited supply chain [90,100] (0)

[…,…]Limits to practice values (…) rate of change of practices per time step

1 Particle emissions are approximately linearly related to SOx emissions

2 Köhler et al. 2018 as source; special procedures required for bunkering

3 Methanol is poisonous, requires extra safety measures vs. HFO (DNV-GL 2014)

4 Liquid, requires minor modifications to HFO bunkering and storage (DNV-GL 2014)

5 Hydrogen requires a different infrastructure to HFO/MGO (DNV-GL 2014)

6 Biofuels and PtL fuels have similar costs (in 2018), of the order of 2-4* MGO (Transformation Bio Project workshop)

Table 2 Initial practices of shipping companies

	GHG Emission CO_2Equiv/MW	Fuel Cost $/Tonne HFO/MGO	Opspeed Av. Kts	NOx SOx emissions % of diesel	CAPEX $/MW	Operational and technology change	Bunkers - Compatible with current HFO/MGO
Mainstream: as diesel at least initially							
Cruise, ferry, RoPax	90 [10,100] (0)	40 [0,100] (0.5)	50=25kts [0,100] (0)	100 [0,100] (0)	40 [30,60] (0)	100 [50,100] (0)	100 [90,100] (0)
solid and liquid bulk including chemicals and Gas carriers	90 [10,100] (0)	40 [0,100] (0.5)	50=15kts [0,100] (0)	100 [0,100] (0)	40 [30,60] (0)	100 [50,100] (0)	100 [90,100] (0)
Container, RoRo	90 [10,100] (0)	40 [0,100] (0.5)	50=25kts 10000TEU [0,100] (0)	100 [0,100] (0)	40 [30,60] (0)	100 [50,100] (0)	100 [90,100] (0)
Platform Support Vessels and Offshore vessels; tugs dredgers etc;, specialist vessels	90 [10,100] (0)	40 [0,100] (0.5)	50 [0,100] (0)	100 [0,100] (0)	40 [30,60] (0)	100 [50,100] (0)	100 [90,100] (0)
NICHES							
Cruise LNG	90 [10,100] (0)	45 [0,100] (0.5)	50 [0,100] (0)	10 [0,100] (0)	40 [30,60] (0)	90 [50,100] (0)	60 [50,100] (0)
Ferry LNG	90 [10,100] (0)	45 [0,100] (0.5)	50 [0,100] (0)	10 [0,100] (0)	40 [30,60] (0)	90 [50,100] (0)	60 [50,100] (0)
PSV_Service LNG	90 [10,100] (0)	45 [0,100] (0.5)	50 [0,100] (0)	10 [0,100] (0)	40 [30,60] (0)	90 [50,100] (0)	60 [90,100] (0)
Wind assist bulkers	80 [10,100] (0)	35 [0,100] (0.5)	50 [0,100] (0)	100 [0,100] (0)	45 [30,60] (0)	90 [50,100] (0)	100 [90,100] (0)

FINANCING SHIPS WITH WIND-ASSISTED PROPULSION TECHNOLOGIES

O. Schinas* and **D. Metzger*** **, *Hamburg School of Business Administration, Germany; **Helmut Schmidt University, Germany

1. SUMMARY

The issue of financing green ships as well as of financing decarbonization of maritime operations has already attracted the interest of experts from the academia and the industry. Given that green ships do not enjoy a freight premium and their operating expenses are at the level of conventional assets, the inherently higher capital expenses required for green assets deteriorate profit margins and deter investors from relevant outlays. Hence, financial engineering is required in order to make investors indifferent to cost and riveted to the decarbonization aspects of any ship financing project. This paper summarizes research results and presents illustrative cases and structures. The analysis considers sharing economy models, such as the 'pay as you save', that perfectly suit solutions involving wind-assisted propulsion, and will also identify risks and ways of mitigating them. The approach considered may assist in the analysis of various wind-assisted and power-boosting technologies.

2. INTRODUCTION

The decarbonization of marine operations and generally the abatement of Green-House Gases (GHG) is currently in progress, as the International Maritime Organization (IMO) adopted the Energy Efficiency Index (EEDI) and considers further measures to reduce CO_2 emissions [10, 7, 8]. The policy aspects of this issue are well analyzed in the academic literature as well as in other sources.

Nevertheless, the issue of decarbonization is complex and intertwined with the abatement of non-GHG, such as of sulfur SO_x and nitrogen NO_x oxides. An illustrative example is the use of liquified natural gas (LNG) as marine fuel. In principle, the CO_2 emissions from LNG-burning engines are about 25% lower in comparison to engines burning Heavy Fuel Oil (HFO), yet methane CH_4 might also be emitted. Methane is a 28 times more potent GHG than CO_2, so in a total life cycle assessment the externalities from the consumption of LNG might be higher, unless methane is trapped and treated properly [12]. In the near future, pollutants such as CH_4 and ammonia (NH_3) are expected to attract the attention of policymakers and regulators, as of the European Parliament [4, 5], which gives a clear sign of global and regional interest in promoting stricter environmental legislation. Thus, innovative technologies would gain importance.

As there is no 'one-size-fits-all' solution, various technologies have been developed in order to reduce consumption and generally to lower energy requirements per unit of transport. Wind-assisted technologies, such as those promoted by the International Wind-Ship Association (IWSA), could offer feasible operational options subject to the limitations of the ship type, size, age, design and main particulars. The key benefit is the right-shift of the resistance curve, i.e. a reduction of the energy required for a given speed. In this regard, the numerical examples considered below are based on Flettner rotors technology as well as on ventifoils. The analysis will consider financing models recently published in academic journals as well as real-world business cases.

Regardless of the selected technology, the financial burden of installing and operating wind-assisted power-boosting equipment boils down to the same set of decision parameters:

- Currently, there is no freight premium offered in the market for greener operators; any benefit is still reflected in the agreed speed and consumption terms in the time-charter markets, or in the freight offer in the spot markets. Preferences of shippers might result in a bias for operators or ships but is not expressed with a premium. The need of deeming zero-emission ships more commercially attractive is highlighted at the highest possible levels; these ships serve the 2050 GHG goals set by the IMO [9] that will not be achieved by only using fossil fuels [16].

- Operating expenses (OPEX) for both conventional and new green tonnage are practically the same. OPEX depend on the type and size of the ship as well as on the Flag and other commercial parameters, that can be omitted in this analysis, in order to compare similar cases. In the following analysis, the letters c and n denote these two classes of ships.

- Capital expenses (CAPEX) are expected to be higher for the new green ships as innovative technologies come with a premium or a risk. Ceteris paribus, a wind-assisted power-boosting technology is an extra detachable or a fixed piece of equipment on a conventional vessel. Therefore, there is a premium for the acquisition, installation and certification of this equipment onboard.

- Voyage expenses (VOYEX) are expected to be lower for the new green ships. Installed technology reduces the energy requirements per unit and therefore all associated costs.

Given the above, the margin between revenues and the sum of the expenses determine the financial attractiveness of the relevant investment for a ship with new technology. The higher the margin, the higher the financial resilience. Assuming a rational decision-maker, who will opt for the solution with higher margins and, in case of similar margins (indifference), opt for the greener solution, as it contains better attributes other than cost. Assuming that revenues and the sum of expenses on a daily basis are equal, then Equations 1–5 determine the necessary conditions for attracting the interest of investors:

$$CAPEX_c + OPEX_c + VOYEX_c = CAPEX_n + OPEX_n + VOYEX_n \quad (1)$$

However

$$OPEX_c = OPEX_n \quad (2)$$

$$CAPEX_n \geq CAPEX_c \quad (3)$$

$$VOYEX_c \geq VOYEX_n \quad (4)$$

Therefore

$$\begin{aligned} CAPEX_c + VOYEX_c &= CAPEX_n + VOYEX_n \\ CAPEX_n - CAPEX_c &= VOYEX_c - VOYEX_n \\ \Delta CAPEX_{nc} &= \Delta VOYEX_{cn} \end{aligned} \quad (5)$$

In other words, the burden in capital expenses should be compensated by lower voyage expenses. Ideally $\Delta CAPEX_{nc} \leq \Delta VOYEX_{cn}$, where the benefit in the operations is larger than the acquisition burden.

3. LITERATURE REVIEW

The focus on financing environmentally enhanced vessels recently attracted the interest of the research community. Ross and Schinas [15], provided empirical evidence that greener ships enjoy higher market valuation, i.e. they have a competitive financial advantage over less green assets. In this regard, the premium of innovation, namely the higher acquisition price of a greener ship is already identified in the literature as a critical decision parameter, along with the weighted average cost of capital (WACC) of the project [19, 17]. This premium is approached similarly to Equations 1-5 above:

$$\Delta PP = PP_n - PP_c = (RFR_n - RFR_c) \cdot C \cdot D_{empl} \cdot \sum_{t=1}^{N} \frac{1}{(r+1)^t} \quad (6)$$

Equation 6 illustrates the relationship of the differential of purchase price PP between a new 'green' ship and a conventional one with the required freight RFR, the commercially useful capacity C and the days of employment D_{empl}; the right part of the equation is discounted over r, a parameter equal to the WACC or to inflation or generally a factor that reflects the impact of time. Equation 6 implies that the market should be willing to accept a higher RFR, namely $RFR_n \geq RFR_c$, as a necessary condition to justify the ΔPP, the purchase premium, otherwise there is no rationale for investing in the new ship [17].

Assuming that r, the WACC of the project, is not identical for conventional and new green ships, as suggested in the formula, then an adjustment of the WACC for the new green ships, could have the same effect as the difference of the RFR. WACC is identified as an important element for the financing of green ships in the industry, and sourcing-based export finance models (SBEFM) are developed that consider the necessary condition $WACC_n \leq WACC_c$. Usually, SBEFM combine guarantees, which are granted by Export Credit Agencies (ExCrA) of countries that are benefitting from the innovative engineering and technology of green ships at the macroeconomic level, with debt and equity [19]. Empirical data further supports the analysis and the arguments for the use of export credit facilities (ECF) are stated. The enhancement of financing possibilities and combinations with debt or equity is highlighted. Nevertheless, skepticism is introduced, as it is questionable whether the magnitude in WACC reduction for the green ship would be sufficient through the SBEFM on a stand-alone basis. Briefly it is shown that ECF are useful but not indispensable.

The analysis is not concluded, unless the issues of the residual value and of the aftermarket of green and of conventional designs are considered. The limited aftermarket could negatively impact the initial acquisition decision, while the residual value, at large linked to the lightship mass of the vessel and strongly depended on market conditions, affects the original calculations and financial planning [19, 17].

The presentation of the literature is not exhaustive; the impact of some of the above parameters in the financial planning and performance of the ships will be discussed in coming sections. This brief overview of the literature concludes with the 'pay-as-you-save' (PAYS) model. This sharing economy model is mainly applicable for wind-assisted power boosting equipment and splits the risks, and revenues based on fuel savings and carbon price between the owner/operator and the supplier/financier. This work introduces a financing model for the installation of greening technology that simultaneously addresses the need of reducing emissions and compensating suppliers and users. This novel approach for a shared savings enables the development of business models fostering operational pragmatism [18].

Apart from the above contributions, this work considers also input from other sources and researchers. The research of Bouman et al. [2] assesses the emission reduction potential of different greening technologies. Their results indicate that wind propulsion technologies have the potential to reduce the carbon emissions of certain ships of up to 50%. However, shipping companies seem to choose suboptimal measures to go green, as Rehmatulla et al. [14] argue. In particular, they name Flettner Rotors as a promising technology, which nonetheless is only implemented on a tiny scale. The financing problem, which is illustrated in the previous sections, might be one reason for neglecting wind propulsion technologies when it comes to greening. Another reason are the low oil prices from 2014 onwards, as argued in several papers such as Atari et al. [1], Zis et al. [21], and Zis et al. [20]. When oil prices increase, costly conventional fuels such as Marine Gasoil (MGO) become unattractive as a greening measure. Consequently, the demand for greening technologies and alternative fuels would rise in the future. In order to overcome financing issues related to greening technologies, researchers such as Halff et al. [6], Karslen et al. [13], and Zis et al. [21] propose subsidies or other kind of external financial aid. However, the shared savings financing schemes may be an alternative to such aid. Those schemes allow shipping companies to reduce the financial burden in regard to capital-binding technologies to a minimum. In return, suppliers or any other financier, who carry the majority of the upfront costs, participate in the money savings achieved by the respective technology. Thus, profits and risks are shared between supplier/ financier and shipping company. The following sections illustrate how a shared savings scheme could look like for a detachable sail and a Flettner Rotor. Therefore, the pay-as-you-save model (PAYS), which is a shipping-specific shared savings model developed by Schinas and Metzger [18], is considered.

4. FINANCIAL PERFORMANCE AND OPERATIONAL CONSIDERATIONS

The financial performance of a ship depends mainly on capacity, speed and consumption as well as on the operational profile, i.e. utilization, routing, port times and congestion, etc. In this regard, wind-assisted power boosting technolo-gies are lowering the consumption for a given speed. The decrease in energy requirements results in less emitted CO_2. Therefore, the benefit from installing such technologies is twin fold: less fuel is consumed, therefore the ship becomes more competitive, as well as less carbon is emitted, and operations become greener. Considering that market-based measures (MBM) will be deployed in the near future in order to motivate further the decarbonization of marine operations [11], this benefit will be further monetized

From an operational point of view, the performance and therefore the monetized benefit of these technologies de-pends on the following parameters:

- wind force;
- relative wind direction - angle of attack;
- clear flow of wind around the device/apparatus;
- aerodynamic performance of the device/apparatus.

Moreover, sound engineering and seamanship require attention to:

- stability;
- local strength;
- local protection, e.g insulation, painting, corrosion;
- cabling for operation and control;
- required energy;
- obstruction of the visibility from the bridge;
- obstruction of cargo handling operations, e.g. interference with the operation of cranes and hatches.

Last but not least, it is possible that a mix of technologies, such as of sails and solar panels, might propel the ship at a relatively low speed and at the same time cover merely or fully its energy requirements for auxiliary systems. This is possible in special designs and operational patterns, such as of Quadriga [3].

The shared-savings financial plan implies that the owners do not bear all the risks; risks are shared on a basis of ratio or a formula between the owner and the supplier of the technology or a financier, who provides the funding. In most cases, equipment providers require a full payment upon installation; as this is not always possible, a financial house or a relevant financial entity intermediates and undertakes the risks and benefits on behalf of the supplier. The financial planning that supports the decision-making process is affected by:

- the sharing ratio, i.e. the split of benefits, such as monetized fuel savings and carbon emissions;
- fuel prices;
- carbon prices (when CO_2 emissions get monetized);
- utilization of the installed power-boosting equipment (and/or its availability in general);
- financial parameters of the power-boosting system, such as;
  - the cost of the equipment including the cost of installation;
  - the necessary first sum for the acquisition and installation (first payment);
  - the economic and/or technical life of the equipment depending on the case;
  - the expected maintenance outlay;
  - the WACC of the project, that depends on the financial standing of the company and/or the financial scheme agreed with the supplier or the financiers.

Given the above, the following numerical examples aim to illustrate the impact of these parameters. The examples are based on real-world cases; the assumptions are made for a detachable ventifoil sail system that is mounted on a 8,000DWT general cargo ship, which operates on about 300 sea days per year:

- The ventifoil sails cost \$500,000 and can be used on 50% of all sea days;

- The sail saves about 10% of fuel consumption per day it is used;
- The ship owner pays 25% of the price (= \$125,000), while the rest is financed by the financier/ supplier;
- 90% of the savings during the first 6 years go to the financier/ supplier;
- The technology can be used for 9 years;
- The IFO180 demand of the ship per sea day is $20t$, which would be typical for a 8,000DWT ship;
- The average price per ton IFO180 is \$300;
- One ton of combusted IFO180 equals 3.114 CO_2;
- In the future, each ton of saved carbon emissions has a value of \$20, which reflects the current IMO climate strategy [9];
- Maintenance costs for year 1: \$5,000 with an annual exponential increase of $e^{0.05}$;
- Weighted average cost of capital (WACC) of the shipping company and discount factor for the financier: 7.0%.

The detailed model with all analytical expressions and clarifications can be found in Schinas and Metzger [18]. Figure 1 visualizes the described example:

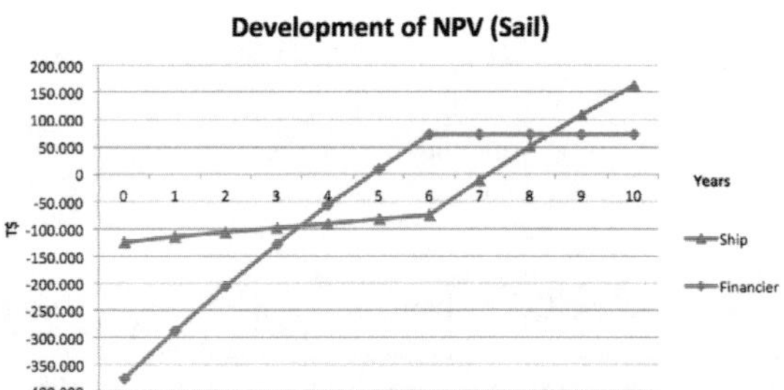

Figure 1: Development of NPV (Ventifoil Sails)

It can be seen that the development of the NPV for both parties changes after the end of the sharing period. After year six, all the savings achieved by the technology are credited to the shipping company's account until the technology cannot be used anymore. Depending on the geographical area in which the ship regularly operates, different utilization scenarios can be calculated (Figure 2).

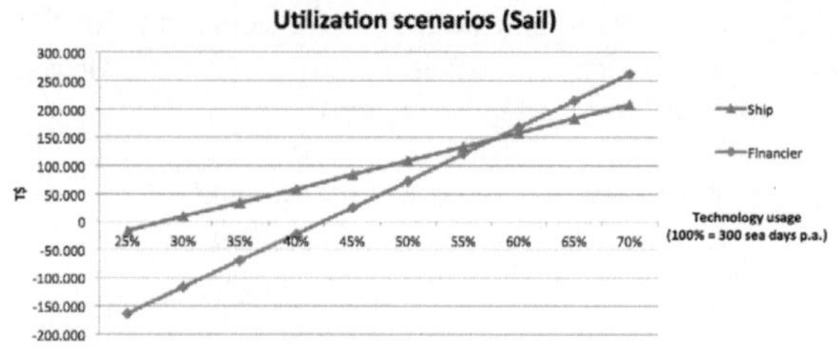

Figure 2: Utilization scenarios (Ventifoil Sails)

Figure 2 shows that -in this example- the supplier/ financier benefits to a higher extent in beneficial circumstances like high utilization rates, but also suffers to a higher extent from unfavorable circumstances like low utilization rates. If the sharing period were shorter or the sharing ratio differed the shipping company's NPV would react more sensitive to such circumstances.

The model is highly parameterized, which allows to figure out optimal conditions for a shared savings contract. E.g. the sharing ratio may be adjusted in order to make the contract more attractive for each party. Figure 3 shows how sensitive the returns of each party react to different sharing ratios.

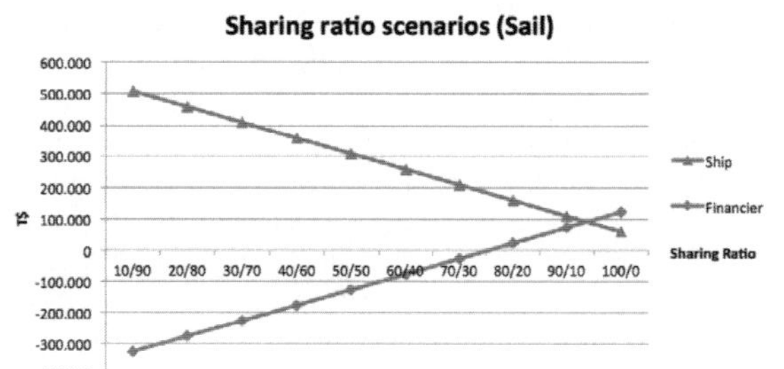

Figure 3: Sharing ratio scenarios (Ventifoil Sails)

Considering the installation of a Flettner rotor solution, it is assumed that the technology in mounted on a ship with the same characteristics as above. The following parameters change:

- A rather sizable Flettner Rotor costs \$1,000,000 and can be used on 60% of all sea days;
- It saves about 13% of fuel consumption per day it is used;
- The ship owner pays 25% of the price (= \$250,000), while the rest is financed by the financier/ supplier;
- 90% of the savings during the first 7 years go to the financier/ supplier;
- The technology can be used for 10 years at least;

- Maintenance costs for year 1: \$10,000 with an annual exponential increase of $e^{0.05}$.

Since the Flettner Rotor requires higher initial cash commitment and reduces the fuel consumption to a higher extent, it is more sensitive to external factors. This is illustrated in Figure 4, which shows several fuel price scenarios. In general, it is reasonable to expect fuel prices to impact the NPV of a greening technology substantially; this is also highlighted in the literature [20].

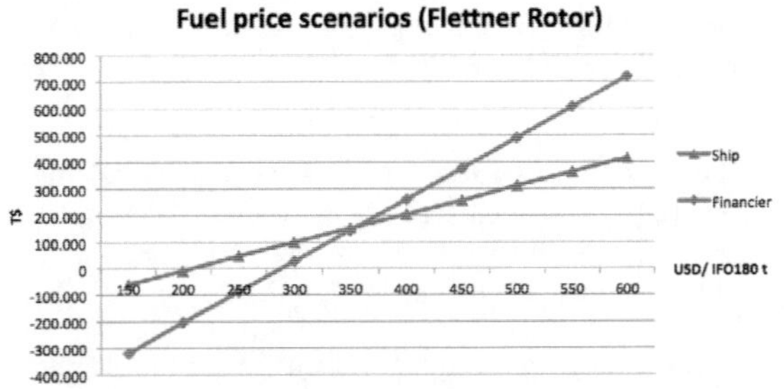

Figure 4: Fuel price scenarios (Flettner Rotor)

Under consideration of the current development regarding the IMO greenhouse gas (GHG) emission reduction strategy [9], the value per ton carbon offset is difficult to determine. Today, it is not even sure whether there will be an emission trading system (ETS) or a levy on GHG emissions. However, the commoditizing of GHG emissions in the foreseeable future is more than probable. Therefore, the base case scenario assumes a value of \$20 per ton carbon offset. Figure 5 visualizes how different values impact the NPV of a Flettner Rotor.

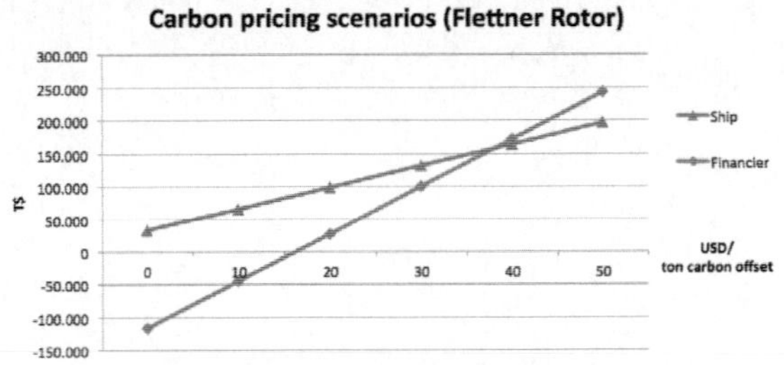

Figure 5: Carbon pricing scenarios (Flettner Rotor)

At this point, it must be said that there is a lack of sufficient performance data for both discussed technologies. Therefore, the assumed values may vary substantially from actual ones. The impact of factors such as geographical area of operation and cargo types on the respective technology's performance is subject to further research.

5. CONCLUSION

As the need of decarbonizing marine operation is growing and further institutionalized, the potential of wind-assisted power boosting systems is growing. These systems can merely or fully cover the propulsion needs of the ship with a zero or limited carbon footprint. Current technical solutions based on fossil fuels cannot serve the goals set. Therefore, wind assisted systems deserve further attention and investigation.

The financing of innovation is identified as a critical element in this process, and research suggests -as necessary but not always sufficient condition- either a premium in the RFR or a reduction of the WACC. Should this be the case, then the premium of innovation, i.e. the higher acquisition cost, is compensated from lower voyage costs, making investors indifferent in terms of cost and motivating them to finance greener ships.

Recent research delivered a sharing economy model, where suppliers of the equipment and owners, operators of the equipment, share the risks and the benefits. In this business model, the benefits of reduced consumption and of carbon emissions are monetized and split according to a transparent and pre-agreed formula. Thus, operators have an interest of selecting operating profiles that increase the benefit and suppliers have an interest in further supporting and upgrading the technology installed, such as critical sensors and software. Numerical examples based on real-world cases considered in the previous section, illustrate the potential of this business model, while sensitivity analysis identifies critical parameters, such as the fuel and carbon price, as well as determine the limits for detailed commercial and technical acceptance.

Apparently, advanced and innovative technical solutions require also financial innovation, so financial engineering is expected to support all processes of decarbonization, and further research and disruption is anticipated. Should the industry fail to enter an upward spiral of innovation that enhances synergies with other industries and technologies towards decarbonization, the introvert and rather ineffective experiences from the abatement of non-GHG might be replicated. In the non-GHG case, technologies, such as scrubbers, increase the fuel consumption and therefore the CO_2 emissions, yet are not penalized as long as there is no carbon price. Ships equipped with scrubbers can also operate at a higher speed, should the market demand and the fuel prices permit, ergo carbon emissions increase. Permitting the use of sulfurized and desulfurized fuels results in a patchwork of rules, damaging essentially the requisite of international and leveled regulation. As decarbonization requires global action, a patchwork approach should be eluded by all stakeholders.

6. REFERENCES

[1] S. Atari, Y. Bakkar, E. O. Olaniyi, and G. Prause. Real options analysis of abatement investments for sulphur

emission control compliance. *Entrepreneurship and Sustainability Issues*, 6(3), 2019.

[2] E. A. Bouman, E. Lindstad, A. I. Rialland, and A. H. Strømman. State-of-the-art technologies, measures, and potential for reducing ghg emissions from shipping – a review. *Transportation Research Part D: Transport and Environment*, 52 Part A, 2017.

[3] S. Cargo. Next generation cargo. Website, 2019.

[4] European Parliament. Air quality: Environment meps call for tougher new national caps on pollutants. Press Release, July 2015.

[5] European Parliament. Air quality: Meps approve new national caps on pollutants. Press Release, October 2015.

[6] A. Halff, L. Younes, and T. Boersma. The likely implications of the new imo standards on the shipping industry. *Energy Policy*, 126, 2019.

[7] IMO. 2014 Guidelines on the Method of Calculation of the Attained Energy Efficiency DesignIndex (EEDI) for New Ships. *MEPC.245(66)*, 2014.

[8] IMO. 2014 Guidelines on the Method of Calculation of the Attained Energy Efficiency DesignIndex (EEDI) for New Ships. *MEPC.254(67)*, 2014.

[9] IMO. Initial imo strategy on reduction of ghg emissions from ships. *MEPC.304(72)*, 2018.

[10] IMO. Prevention of air pollution from ships. Press Release, 2018.

[11] IMO. Market-based measures. Press Release, 2018.

[12] IPCC. Fifth assessment report (ar5). Technical report, 2015.

[13] R. Karslen, G. Papachristos, and N. Rehmatulla. An agent-based model of climate-energy policies to promote wind propulsion technology in shipping. *Environmental Innovation and Societal Transitions*, 2019.

[14] N. Rehmatulla, J. Calleya, and T. Smith. The implementation of technical energy efficiency and co2 emission reduction measures in shipping. *Ocean Engineering*, 139, 2017.

[15] H. Ross and O. Schinas. Empirical evidence of the interplay of energy performance and the value of ships. *Ocean Engineering*, in print, 2019.

[16] Safety4Sea. Imo: Zero-emission ships should be more commercially attractive. Press Release, 2019.

[17] O. Schinas. Chapter 8 - financing ships of innovative technology. In S. Gong and K. Cullinane, editors, *Finance and Risk Management for International Logistics and the Supply Chain*. 2018.

[18] O. Schinas and D. Metzger. A pay-as-you-save model for the promotion of greening technologies in shipping. *Transportation Research Part D: Transport and Environment*, 69, 2019.

[19] O. Schinas, H. H. Ross, and T. D. Rossol. Financing green ships through export credit schemes. *Transportation Research Part D: Transport and Environment*, 65, 2018.

[20] T. P. V. Zis, P. Angeloudis, M. G. H. Bell, and H. N. Psaraftis. Payback period for emissions abatement alternatives: Role of regulation and fuel prices. *Transportation Research Record Journal of the Transportation Research Board*, 2549(1), 2016.

[21] T. P. V. Zis, H. N. Psaraftis, G. Panagakos, and J. Kronbak. Policy measures to avert possible modal shifts caused by sulphur regulation in the european ro-ro sector. *Transportation Research Part D: Transport and Environment*, 70, 2019.

7. AUTHORS' BIOGRAPHIES

Orestis Schinas is the Professor of Shipping & Ship Finance and Head of the Maritime Business School at the Hamburg School of Business Administration (HSBA). His professional career includes large corporate projects, such as IPO and advanced research and business development assignments in the finance, shipping and port sectors and has provided his services to private and public entities in Europe, Africa and the GCC, as well as to international organizations. Schinas' academic work focuses on the commercial and financial aspects of maritime ventures while he retains a keen interest on sustainable business development.

Daniel Metzger is a doctoral student at the Hamburg School of Business Administration (HSBA) and the Helmut Schmidt University (HSU). He researches in the field of green shipping with a focus on financial and policy aspects, and has also gained substantial professional experiences in the financial sector.

RETROFITTING OF FLETTNER ROTORS – RESULTS FROM SEA TRIALS OF THE GENERAL CARGO SHIP "FEHN POLLUX"

M. Vahs, Hochschule Emden-Leer - University of Applied Sciences, Germany

SUMMARY

In 2018, a German- Dutch project consortium under the scientific direction of the Emden/Leer University of Applied Sciences retrofitted and commissioned the latest rotor development of the Eco-Flettner type on the test ship *Fehn Pollux* of the Leer-based shipping company Fehn Ship Management. The retrofitting concept is groundbreaking in terms of easy transferability to other ships. Upscaling to a significant share of the world merchant fleet could make a substantial contribution to climate protection.

The most frequently asked question in connection with modern sail drives is about the performance potential and the associated fuel savings. Transparent performance data is required to enable an economic prognosis for the use of Flettner rotors on ships. The Faculty of Maritime Sciences at Emden/ Leer University of Applied Sciences has developed an automatic control and monitoring system for Flettner rotors that also records extensive operating and environmental data.

The data shows that all previous assumptions and model calculations are basically correct. With regard to the performance potential, the first series of measurements show even higher rotor forces compared with model calculations. This is a further benefit for Flettner rotor efficiency and could help the technology achieve a breakthrough as a building block for low-emission shipping.

Figure 1: *Fehn Pollux* outbound for sea trials (Source: MariGREEN)

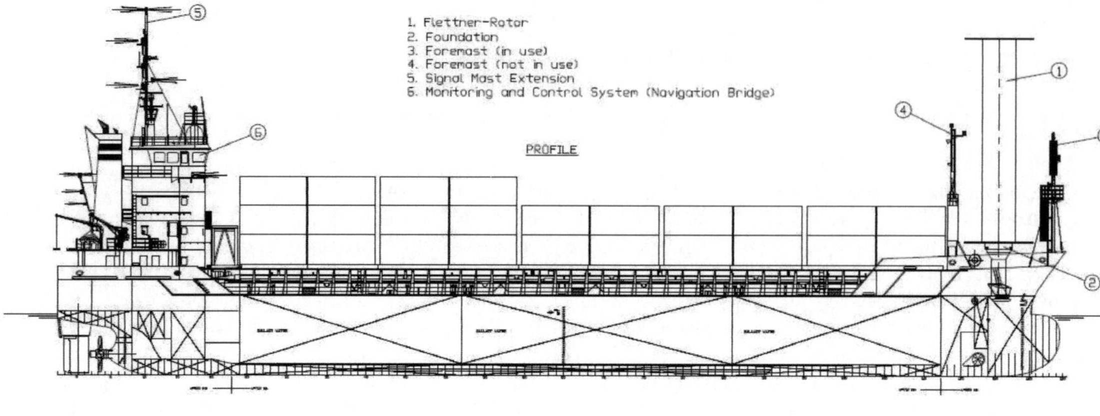

Figure 2: Arrangement drawing for retrofitting *Fehn Pollux* (Source: ABH Ingenieurtechnik GmbH, Emden)

NOMENCLATURE

[Symbol]	[Definition] [(Unit)]
AWA	Apparent Wind Angle (deg)
AWS	Apparent Wind Speed (kn)
C_L, C_D	Lift and Drag Coefficient [-]
C_X, C_Y	Coefficients for forces in ship fixed x- and y-direction [-]
F_X, F_Y	Forces in ship fixed x- and y-direction [kN]
PB	Break power of ship's main engine (kW)
RPM	rotational speed (revolutions per minute)
SOG	Speed over ground (kn)
TWA	True Wind Angle (deg)
TWS	True Wind Speed (kn)

1. INTRODUCTION

The IMO aims to reduce CO2 emissions of international shipping by at least 50 percent by 2050. In this way, shipping would make a significant contribution to the climate target, even if the IMO remains below the demands of the high ambitious states that want to achieve 70 to 100 percent decarbonization in this period. Even though many details of the new IMO strategy are still open, it is already clear that new technological approaches are needed. A renaissance of wind energy for ship propulsion seems to be obvious, the high potential at sea is well known. Today's shipping needs technologies that meet all the important requirements of modern ship operation. Flettner rotors have already proven their fundamental suitability in various projects. In June 2018, a German-Dutch project consortium under the scientific direction of the Emden/Leer University of Applied Sciences retrofitted and commissioned the latest rotor development of the "Eco-Flettner" type on the test ship *Fehn Pollux* of the Leer-based shipping company Fehn Ship Management. The retrofitting concept developed by the maritime companies and scientific institutions involved is groundbreaking in terms of easy transferability to other ships. Upscaling to a significant share of the world merchant fleet could make a substantial contribution to climate protection.

The following report is intended to provide information on important aspects of planning retrofits or newbuildings with Flettner rotors and is based on the results of the sea trial and the first phase of testing under normal operating conditions. In order to comply with all existing regulations regarding the use of this new technology, various areas of ship technology and navigation had to be illuminated, e.g. proof of safe manoeuvring characteristics and sufficient stability. The most frequently asked question in connection with modern sail drives, however, is about the performance potential and the associated fuel savings. Transparent performance data are required to enable an economic prognosis for the use of Flettner rotors on ships. The Department of Maritime Sciences at Emden/Leer University of Applied Sciences has developed an automatic control and monitoring system for Flettner rotors that also records extensive operating and environmental data. The data show that all previous assumptions and model calculations are basically confirmed. With regard to the performance potential, the first series of measurements show even higher rotor forces compared to the model calculations. This would have a positive effect on the economic efficiency of Flettner rotors and could help the technology to achieve a breakthrough as a building block for low-emission shipping.

2. RETROFIT CONCEPT

An important development objective for the Eco-Flettner within the framework of the "MariGreen"[1] project is the transferability of the retrofitting concept to a larger share of the existing world merchant fleet in order to achieve significant reductions in fuel consumption and emissions. The structural features of the multipurpose freighter *Fehn Pollux* provide an ideal basis for this, as there are a large number of ships worldwide with a similar arrangement of holds and superstructures.

By installing a wind propulsion system supporting the main engine, the potential for average fuel savings in the order of 10 percent was to be demonstrated on the test ship. In previous analyses, the Flettner rotor was technologically favoured because it combines high sailing performance with minimum space requirements and the advantages of a fully automated system. Furthermore, the construction is robust and insensitive to wear, which offers further advantages compared to other sailing systems, e.g. based on textile sails adapted from the field of yachting.
The central question of the choice of the installation location on the ship was preceded by a detailed analysis of all relevant factors. In order to avoid an impairment of the existing operating procedures, a rotor position outside the cargo area had to be found. The only installation site available was the foreship in front of the cargo holds. The aerodynamic and hydrodynamic conditions were investigated using CFD flow simulations and model tests. The installation on a raised foundation and the use of a lower end plate not provided for in previous Flettner rotors ensure very favourable flow conditions at the rotor. The introduction of the sail forces in the forecastle deck leads to a reduction of yaw moments caused by side forces of the sail yielding a good steering behaviour of the ship during rotor operation. It is therefore advantageous over a midship or aft installation.

The advantages of the installation on the forecastle deck faced the following problems and questions:

- How can the rotor be protected from wave impact?
- How does the restriction of the view from the bridge and the radar performance affect nautical operation?
- What influences on the manoeuvring behaviour can be expected from the rotor forces?

The rotor was reinforced in accordance with existing classification regulations to counteract the effects of wave impact, which led to a slight increase in the weight of the rotor. The DNVGL has checked and approved the compliance with the regulations regarding the visibility from the bridge and the radar visibility in advance.

Further investigations by the Department of Maritime Sciences of the University of Applied Sciences Emden/Leer included simulations on the full mission shiphandling simulator, the results of which were validated during the sea trial. The fulfilment of all regulatory requirements for manoeuvring behaviour was also investigated in a study on the shiphandling simulator and verified during the sea trial. Details are given below.

3. NAVIGATION ASPECTS

3.1 VIEW FROM THE BRIDGE

In the first step, arrangement plans had to be submitted to demonstrate that both the view from the bridge and the operating conditions of the radar meet the requirements of relevant regulations.[2] The rotor causes a visually blind sector of 2 degrees in the area of the midship line and thus remains clearly below the limit value of 5 degrees. To check the blind sector, the officer of the watch must change the position on the bridge at regular intervals. During the installation works, a special training programme was conducted for the staff including exercises on a full mission shiphandling simulator. A validation took place during the sea trials. The effect of the rotor on bridge visibility was not perceived as significant by the nautical experts involved in the test and is similar to that on ships with cranes in the midship line, as shown in Figure 3. Changing the position on the bridge by a few metres allows an unrestricted field of vision.

Figure 3: View from the bridge (source: Hochschule Emden/Leer, University of Applied Sciences)

3.2 RADAR DETECTION

During planning of the rotor installation, a blind radar sector behind the rotor in the size of one degree was assumed according to verified plans. Due to the different locations of the two radar antennas, however, there is an unrestricted radar view when both antennas are used simultaneously. However, the blind sector could not be

[1] www.marigreen.eu

[2] e.g. SOLAS Chapter V, Regulation 22

observed during the sea trials. A radar detection test was carried out with a small vehicle. Figure 4 shows that the vehicle could be observed on the radar at close distance in front of the ship, although it was completely covered by the rotor. Possible causes for the detection of the target in the blind sector of the rotor are its material properties (GRP) with respect to the transmission of radar waves and diffraction effects in the shadow area.

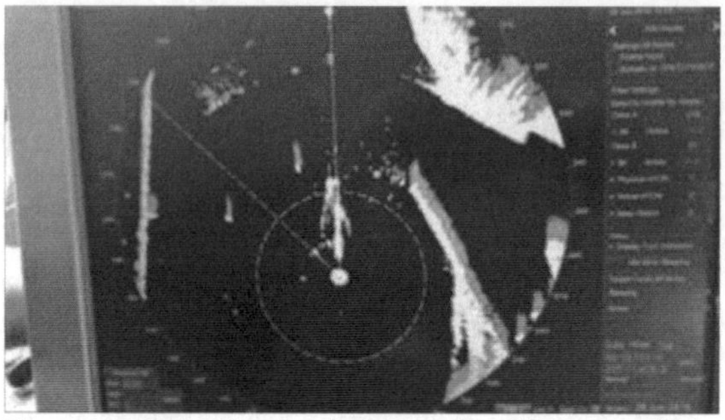

Figure 4: Radar image during a detection test with a small vehicle in the rotor shadow (Source: Hochschule Emden/Leer, University of Applied Sciences)

4. MANOEUVRING BEHAVIOUR

For the sea trials, a manoeuvring programme was drawn up in coordination with the DNVGL classification society to demonstrate conformity with IMO Resolution MSC 137 (76) "Standards for Ship Manoeuvrability" [1] and the DNV-GL guideline 'Certification and Classification Procedures associated with installation of a "Flettner" Rotor Unit onboard a classed vessel' [2]. The main focus was on the influences on the steering and turning characteristics caused by the aerodynamic forces of the rotor. This had already been preceded by an investigation on the "susanNE" ship handling simulator at University of Applied Sciences Emden/Leer. In cooperation with the Leer based company Nautitec, a virtual model of the *Fehn Pollux* for manoeuvring tests on the simulator was developed. This model was used to predict the effects of rotor forces on the ship's manoeuvring behaviour. Furthermore, a "Familiarization Course" lasting several days was developed and conducted for the ship's command. The results of the simulator study by Kramer [3] were validated during the sea trials.

4.1 STEERING PERFORMANCE

The influence on the steering characteristics was tested with maximum transverse forces of the rotor on courses in the wind. During the test runs, approx. 40 kN side force was achieved with moderate winds of strength 4, which corresponds to approx. 50% of the maximum side force. The ship could easily be kept on course with autopilot control. The required rudder angles were below 5 degrees at a ship speed of approx. 10 knots, and below 10 degrees at slow speed of approx. 5 knots. The small effect on the steering capability of the test ship can be attributed to good rudder performance values and, in particular, to the fore ship installation close to the hydrodynamic centre of effort of the hull.

4.2 TURNING CIRCLE

Compliance with the minimum values for Advance and Tactical Diameter in the turning circle was checked by comparative turning circles with the rotor stopped and running at moderate wind speeds (Beaufort 4). The differences were relatively small and showed a good alignment with the values determined beforehand on the simulator. The actually obtained values were far below the permitted minimum values of IMO Resolution MSC 137 (76) for the Advance of 4.5 times the ship's length and for the Tactical Diameter of 5 times the ship's length and should not be regarded as critical at higher wind speeds. The installation located on the fore ship also has a favourable effect here, as the yaw moments caused by rotor side forces are comparatively smaller to an installation further aft in the ship.

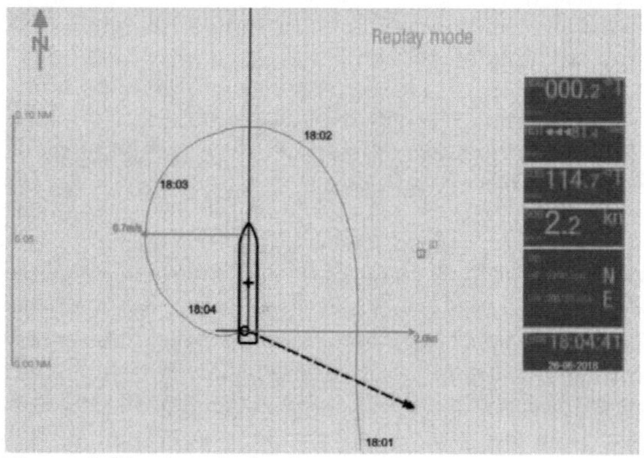

Figure 5: Turning circle plot

4.3 STOP FUNCTION OF THE ROTOR

In order to stop the effect of the rotor forces on the ship as quickly as possible, e.g. in safety-relevant situations, an electrically acting brake was provided. After pressing the stop button on the bridge console, the rotor is stopped within approximately 5 minutes (Figure 6). The stop curve has a linear characteristic. It is recommended to stop the rotor beforehand in order to avoid any influence on the manoeuvring behaviour when carrying out port manoeuvres.

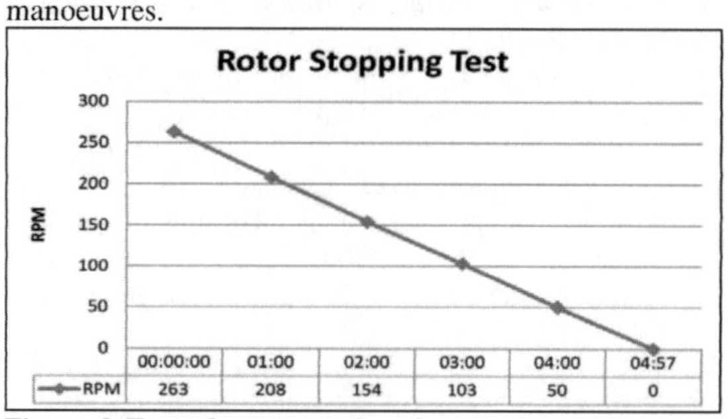

Figure 6: Test of rotor stop function

5. ROTOR PERFORMANCE

5.1 MODELING AND SIMULATION

For the prediction of the rotor's propulsion power and associated fuel savings, numerous model tests and simulations were performed using the wind tunnel of TU Hamburg-Harburg and numerical computational fluid dynamics (CFD) methods at the Center for Modeling and Simulation at Emden/Leer University of Applied Sciences [4]. The rotor's geometry including end plates was optimized. Lift and drag coefficients from wind tunnel model testing are presented in Figure 7.

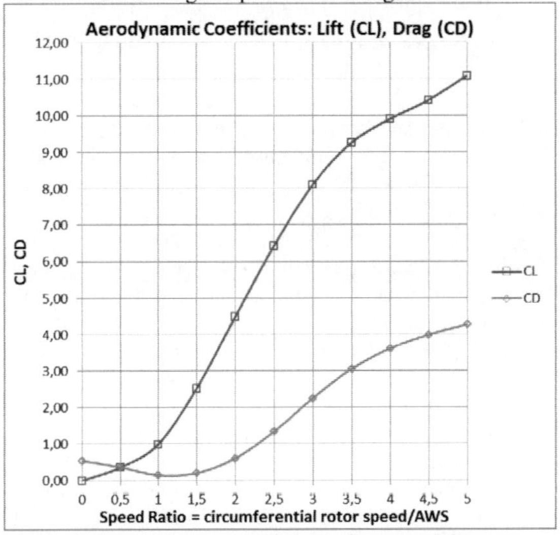

Figure 7: Lift and drag coefficients from model tests

The rotor's control and monitoring system developed and patented by scientists at the university displays the rotor forces and propulsion power in real time and comprises additional functions for performance optimisation [5]. In addition to values from model calculation, force measurements are displayed in real time. The rotor force is measured in two axes (longitudinal and side force) by sensors specially adapted for onboard application.

5.2 MEASURMENT METHOD

The force measurement sensors were calibrated by means of a tensile test with load cells. Since certain inaccuracies cannot be excluded during the calibration, an additional measuring method for validating the rotor forces under real operating conditions was tested. For this purpose, a high-resolution measurement of the ship's heeling angle resulting from rotor side forces was carried out during the sea trials. The use of this principle is comparable with the well-known inclining test at the shipyard or for the support of ship's cargo operations to determine the initial stability of the ship. However, the test setup was carried out in the opposite way. The initial stability of the ship was determined precisely by a shipyard inclining test before the sea trials. With a known initial metacentric height, the angle of heel caused by rotor side forces was measured. Heeling moment and rotor side force were recalculated from the heeling angle. For the conduct of the test the ship was steered on a course against the wind, so that the rotor side force reaches a maximum and no crosswind influences the measurement. The rotor was switched on and off several times during the test period in order to measure the differential heeling angle generated by the rotor's side force. With almost smooth seas in protected waters only minimal ship's motion superimposed the measuring process, so that a clear result was obtained. As shown in Figure 8, the rotor forces recalculated from the heeling angle show a high correlation with the measured values of the force sensors and the values of the model calculation. The values of the rotor force both from the sensors (red) and from the heeling measurement (blue), however, are on average about 10% to 40% above the values of the model calculation (yellow). From this it can be concluded that a full scale Flettner rotor generates higher aerodynamic forces under real conditions than predicted by model calculations based on small-scale wind tunnel testing. This may be caused by the influence of the ship structures on the air flow, surface effects on the rotor as well as errors from upscaling the model test results. Since this project was the first to perform precise force measurements on Flettner rotors in real ship operation, no comparative results from other projects were available. Meanwhile recent experiments on Flettner rotors at critical and supercritical Reynolds numbers by Bordogna et al. [6] indicate that Flettner rotors are significantly dependent on the Reynolds number which has not been duly regarded in the prediction of full scale rotor performance so far.

5.3 ROTOR PERFORMANCE IN SERVICE

The additional thrust generated by the Flettner rotor can be used either to increase ship's speed for saving voyage time or to reduce power and fuel consumption of the main propulsion system at constant speed. Therefore, in addition to the force measurement on the rotor, the ship's speed was measured with GPS and the fuel quantity was measured by a signal transmitter to record the fuel rack setting on the main engine. Since the electric drive of the Flettner rotor is fed from the shaft generator and thus directly from the main engine, the fuel measurement provides the correct total consumption including rotor operation. To evaluate the rotor performance, the ship's speed and engine power as well as fuel consumption with and without rotor was compared. However, as the ship's speed at given propulsion power depends on many factors, it is difficult to evaluate the data. To solve the problem, measurements were carried out over short periods under constant environmental conditions, during which the Flettner rotor was first switched on and then switched off again. If the main engine setting remained constant, the increase in ship's speed could be measured and a backward calculation of the rotor power could be made. If the vessel achieved the same speed when the rotor was switched off, the influence of environmental

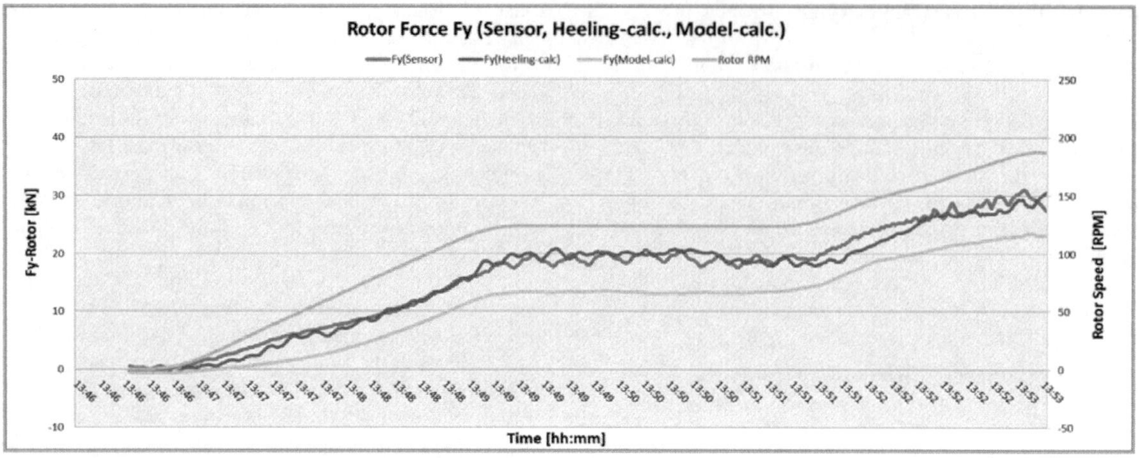

Figure 8: High correlation of rotor force values from force measurement (red), heel angle measurement (blue) and model calculation in real time (yellow). The rotor force is controlled by the rotational speed (green). The measurements show significantly higher values than the model calculation.

factors on the ship's speed during the measurement could be excluded. Similarly, after switching on the Flettner rotor, the fuel rack setting of the main engine could be reduced and measured while maintaining a constant ship speed. The measurement results were used to validate the previous model calculations.

The following example shows the increase and decrease of the ship's speed by switching the rotor on and off with constant main engine setting. The diagram in Figure 10 shows the relationship between rotor's rotational speed (RPM), rotor's propulsive thrust (Fx) and ship speed over ground (SOG). The fuel rack setting of the main engine is initially constant and then reduced at the end of the measurement. The fluctuating fuel rack filling values result from the engine controller keeping constant propeller pitch and RPM under changing load conditions in rough sea. The ship steers a south course (180 degrees) with easterly winds of force 7 on the Beaufort scale (TWS about 30 knots), so the wind direction is from port abeam (TWA approximately 90 degrees, see Figure 9).

The measured values in the diagram show a clear correlation of the rotor thrust curve (Fx, red and yellow) and the ship speed (SOG, blue) with the rotor speed (green). When the rotor accelerates to 260 rpm (100%), the measured rotor thrust values (Fx measured, red) go up to a maximum of approximately 70 kN (70%). The ship's speed increases by 2.5 knots – from 7.6 to 10.1 knots – with almost constant main engine filling (fuel rack, brown).
At full rotor speed the filling is then reduced by approximately 15% with only a minor influence on the ship's speed. When the rotor is switched off, the rotor speed and rotor thrust go back to zero within approximately five minutes, correlating with the reduction of the ship's speed. The measured rotor thrust (Fx measured, red) is significantly higher than the model calculation (Fx model calc, yellow), confirming the results of the sea trials.

To evaluate the rotor power, it can be compared with power from the main engine. In the test, the rotor was used to increase the ship's speed and compared with the equivalent power required for the main engine to achieve the same speed increase. This is based on model tests carried out in the towing tank of the DST research institute at the University of Duisburg-Essen in order to determine the required main engine power as a function of ship's speed. The resulting function curve must be corrected for added resistance and deviating draught/trim that occur under real conditions (service conditions). The measured ship's speed over ground must be corrected for the influence of the prevailing ocean current (here approximately +0.3 knots). Figure 11 shows the red speed-power curve for sailing under service conditions. The function curve is laid through the measuring point of the initial situation "Ship without rotor" (Rotor Off). The equivalent power of the main engine for the same speed increase as produced by the rotor (Rotor On) can be estimated by the red curve. The resulting propulsive power of the rotor is in order of 700 kW of power provided by the main engine, taking efficiencies of gearbox, shaft and propeller into account. This is above the main engine power of approximately 600 kW. In normal charter service of *Fehn Pollux*, the main engine power is limited to about 650 kW (Eco-Speed).

In summary, the following performance figures were recorded from the test:

- Rotor thrust approximately 70 kN,
- Rotor propulsion power approximately 700 kW equivalent to the main engine.

For assessment purposes it should be pointed out that these are high performance values that are close to the performance maximum. However, they are significantly higher than the previous assumptions based on model tests.

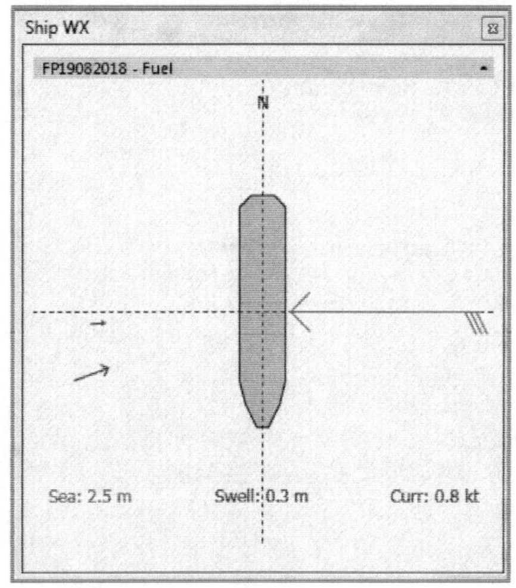

Figure 9: Wind, sea, swell and current conditions at the time of measurement (Source: SPOS MariGREEN-Version, Meteogroup)

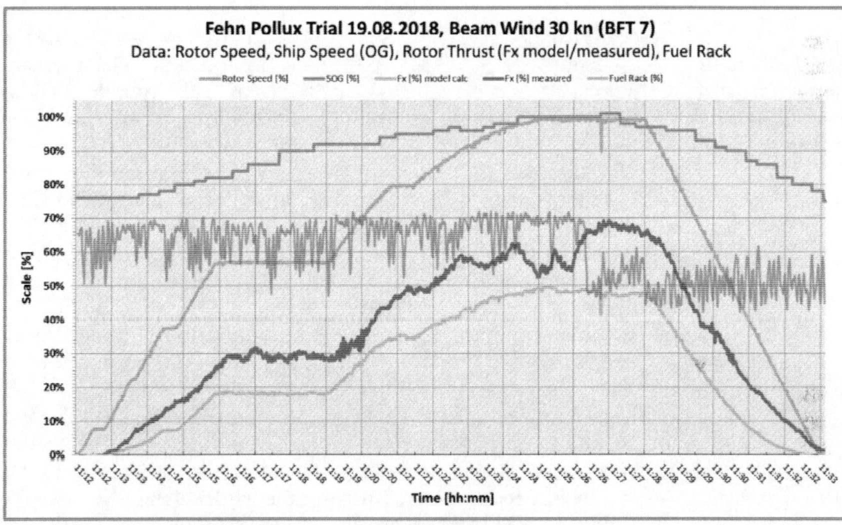

Figure 10: Graph showing the measured values of a performance test during regular service of the Eco-Flettner rotor on *Fehn Pollux*

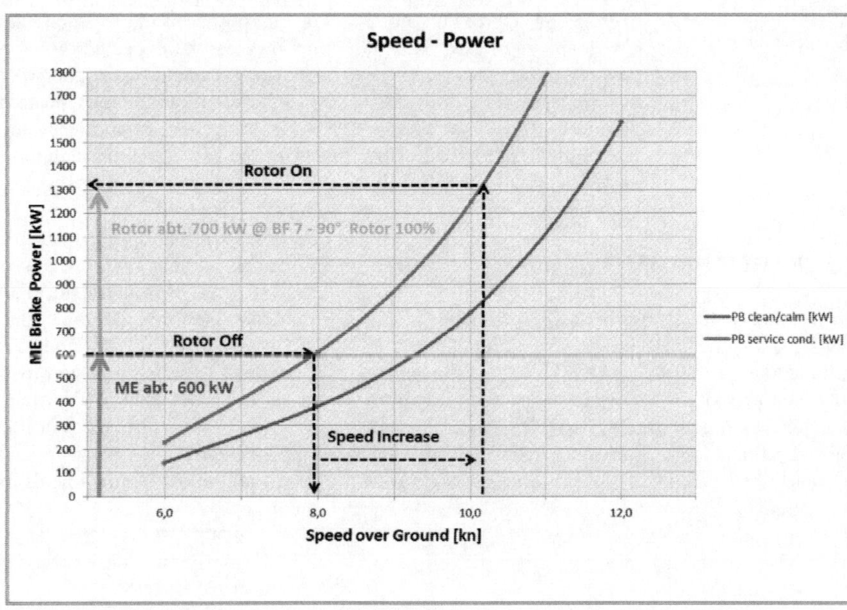

Figure 11: Propulsion power from the rotor and the main engine in relation to ship speed

6. TECHNICAL DATA

Vessel Data *Fehn Pollux*

Type	Multi-purpose freighter suitable for containers and grains
Year of Construction	1996
Length o.a.	89.77 m
Beam	13.17 m
Draught (S)	5.68 m
Deadweight Capacity (S)	4211 t
Tonnage	2844 GT
Main Engine	MWM Deutz SBV 9M 628, 930 kW
Speed	about 10 kn
Rudder	Becker type flap rudder

Rotor Data

Type	Eco-Flettner
Height of the cylinder	18.00 m
Diameter of the cylinder	3.00 m
Diameter of end plates	6.00 m
Projected area	54 m²
Speed	263 RPM max
Drive	E-motor, max 75 kW, average power output depending on the wind conditions, e.g. 30 kW
Thrust	depending on wind conditions, limited to max 80 kN
Propulsive Power	The savings potential depends on the wind conditions along the route and other factors. Under medium-to-good wind conditions, an annual average of approximately 2 kW main engine equivalent power can be saved per 1 m² of projected rotor area (for guidance). Precise predictions are made by route simulations for the specific ship.

7. ACKNOLEDGEMENTS

The development and testing of the "Eco-Flettner" wind drive was part of the MariGreen project, funded within the framework of the INTERREG V A programme Germany-Netherlands with funds from the European Regional Development Fund (ERDF) and through national co-financing from Germany and the Netherlands. Lead partner of the project was MARIKO GmbH in Leer.

The aim of the MariGreen project was to prepare the maritime industry, especially small and medium-sized enterprises, for the future requirements of environmental protection, climate protection and resource and energy efficiency in shipping through cooperation with universities and research institutions. An essential prerequisite for the realisation of the project was the cooperation in the German-Dutch border region in the field of Green Shipping that had developed in recent years.

13 companies and research institutions from Germany and the Netherlands worked together to develop the Eco-Flettner.[3]

8. CONCLUSIONS AND OUTLOOK

The first test results of the *Fehn Pollux* showed that the retrofit concept of a Flettner rotor on the forecastle deck has advantages in terms of aerodynamic behaviour which have a positive effect on the propulsion performance and the manoeuvring characteristics. The disadvantages in terms of visibility from the bridge and radar detection are within acceptable limits and fulfilling all legal requirements. For the forthcoming transition of the world merchant fleet towards sustainable and low carbon marine propulsion it is important that a rotor installation on the foreship is transferable to a significant number of other vessels and can lead to considerable CO2 savings without limiting other ways of saving.

According to validated performance models and route simulations, the annual average power potential of the Flettner rotor installed on the *Fehn Pollux* is in the range of approximately 100 kW to 150 kW in addition to the main engine power. These values naturally depend on the wind conditions along the route. In this case they apply to a trade along the coasts of northern and western Europe. Depending on the required ship speed and the actual power of the main engine, savings of approx. 10% to 20% can be expected on the *Fehn Pollux*. A reduction in speed leads to higher percentage savings, so a combination with slow steaming could lead to higher overall percentage savings. The previous data and results from the project already allow a relatively reliable prognosis of the achievable savings. The recorded data and results from the first project phase already allow a relatively reliable prognosis of the achievable savings. Clear data transparency was provided during the project by validated measurements. This is essential to reassure shipowners, shipbuilders and investors about the validity of this new technology and to enable them to carry out economic appraisals of their own. The project can also be used as proof that converting fleets in this way can reduce CO_2 emissions, but that the process must be supported by appropriate measures in order to achieve important climate policy goals in good time.

---

[3] http://marigreen.eu/projects/wind-windship-engineering-and-design/

When considering the performance values of a Flettner rotor, it must be borne in mind that the installed sail area is decisive for the share of the total power of a ship. For newbuilding, more advanced concepts with better integration of the rotors into the ship's design are conceivable.

9. REFERENCES

1. INTERNATIONAL MARITIME ORGANISATION (IMO), 'Standards for Ship Manoeuvrability', Resolution MSC.137(76), adopted 2002.

2. DNVGL, 'Certification and Classification Procedures associated with installation of a "Flettner" Rotor Unit onboard a classed vessel', Design Guideline, 2016.

3. KRAMER, L., 'Development of the Basic Principles for the Preparation of a Manoeuvring Handbook for Ships using Flettner-Rotors as additional Wind Propulsion System demonstrated by the example of *Fehn Pollux* as Part of the Project MariGreen - Eco Flettner', Bachelor Thesis at Emden/Leer University of Applied Sciences, 2017.

4. STRASSER, S., 'Wind tunnel investigation for the optimized installation of Flettner rotors as wind assisted ship propulsion in the Wind-Hybrid-Coaster project', Bachelor Thesis at Emden/Leer University of Applied Sciences, 2015.

5. MÜLLER, M et al., 'An Intelligent Assistance System for Controlling Wind-Assisted Ship Propulsion Systems', IEEE 17th International Conference on Industrial Informatics (INDIN), 2019.

6. BORDOGNA, G. et al., 'Experiments on a Flettner rotor at critical and supercritical Reynolds numbers', Journal of Wind Engineering & Industrial Aerodynamics, 2019.

10. AUTHORS BIOGRAPHY

Michael Vahs is professor at the Faculty for Maritime Sciences of Hochschule Emden/Leer, University of Applied Sciences, Germany. His research work is strongly related to green shipping subjects including the development and operation of wind propulsion systems for commercial ships. His experience includes a career at sea up to the captain's position and the management of newbuilding and retrofitting projects with the conduct of prototype testing and evaluation.

AUTONOMOUS SAILING VESSELS FOR SHORT SEA SHIPPING

A. Chaplin, OneSails, UK

SUMMARY
This project examined the coast of the UK to find a route to test proof of concept vessels to demonstrate the commercial and functional viability of small Windpowered Autonomous Surface Vessels (WASV) for cargo transport.
Several suitable areas were identified, and one area, the Firth of Clyde, was investigated more closely. Suggestions were limited to currently active ports and cargos investigated were limited to the minor bulk trades. A pair of suitable ports, Campbeltown and Troon, were identified and a cargo, round wood timber, was found.
Over the route identified the costs of using road transport, conventional small coasters and WASV were estimated. It was found that the cost of using the WASV would be comparable with the costs of the conventional alternatives, with the advantage of reduced carbon emissions and potentially less intrusive traffic patterns.

1. INTRODUCTION

The transport infrastructure of Europe is heavily used. Expanding it is expensive, hard and often slow. The carbon emissions associated with transport need to be reduced. Small WASV working as coasters have the potential to reduce carbon emissions and return to commercial use some ports that are too small to be economically served by conventional vessels.

There are many projects that are developing windships of various types and many that are developing autonomous vessels, as can be seen at the annual Autonomous Ship Technology symposium in the Netherlands. There are fewer organisations spotting the possible gains available by linking the two technologies and most of those who are are so far concentrating on the use of their vessels for oceanography and other survey work.

2. THE OVERALL CONCEPT

Short sea and coastal shipping has a long history of providing valuable transport services. In the EU it already carries more cargo than any other mode, bar road, and has the potential to be a significant part of the decarbonisation of transport in and around Europe, but the current fleet is not really fit to play its part in this endeavour. There are many economic and regulatory blocks to the industry really pulling its weight [1], but the ships and ports they use are also part of the problem.

Three of the principal economic drivers in shipping; the capital cost, the crew cost and the fuel cost, push towards larger ships. However for coastal trades the balance between shorter voyage times and unloading time are drivers towards smaller vessels, as smaller vessels spend a smaller proportion of their time static in port being loaded and unloaded and more time moving. The optimum size of a vessel for a particular trade is a function of both the costs of running, loading and unloading the ship, the available consignment size and the distance between the ports. If a way could be found to reduce the financial drivers towards larger ships, coastal shipping could become both more economically attractive and flexible. By using WASV two of these upward drivers are reduced.

Historically the most successful size of sailing coaster has been vessels of about 25m - 30m in length overall carrying in the region of 150 tons of cargo. This is also a size that readily lends itself to technology transfer from the active yachting market. The wage and victualling bill on vessels this size, even if they were allowed to operate with a crew of only two that Thames barges worked with in the early 20th century, would be disproportionate to the cargo carried, rendering them uneconomical when compared with other transport modes. In most cases a conventionally powered vessel of this size would have too high a specific fuel consumption (i.e. would burn too much fuel per tonne of cargo moved) to be commercially viable.

Autonomous shipping will clearly be part of the future of shipping, it was identified by the UK Marine Industries Alliance (UKMIA)[2] as an area of focus for example and autonomous ships are already on the water and being tested and are being first trialled on short sea and coastal routes [3]. Earlier work has identified that one of the most likely obstacles to the adoption of autonomous vessels for international freight transport is the ability of the vessel to respond to machinery failure and other unforeseen events. By choosing coastal routes it gives the ability to easily intervene in such circumstances. If combined with wind power these vessels may be able to offer freight rates comparable to conventional shipping and other transport modes whilst having significantly lower carbon emissions.

The vessels so far envisaged in this study can most easily be described as a descendant of the Thames Barge. As with the Thames Barge the hull would be a very simple hard chine form with a relatively high block coefficient. This is a simple hull form to build, helping to keep the build cost down. It has the added benefit that, due mainly to the form stability the design offers, in most waters, it is possible for the vessel to sail without the necessity of taking on ballast when unladen. This has a cost benefit as it minimises the requirement for ballast tanks and the associated pumps and control systems. Unlike the Thames Barge the vessels would be equipped with bilge keels rather than lee boards. Although leeboards are hydrodynamically more efficient than fixed bilge keels and offer a draft benefit they are more complex mechanically, hence adding to both the capital costs and the ongoing maintenance costs. Having bilge keels and a skeg supported rudder also gives the vessels a naturally stable 'three point landing' if they are required to dry out at a smaller port.

As an auxiliary means of propulsion an electrical motor with sufficient power to achieve a service speed of about 6kts is envisaged. Power for this would come from a bank of batteries which would also provide power for sail handling, bow thruster, rudder control and other such needs. The batteries would be charged from a mixture of sources. When the vessel has spare speed the propeller could be used to harvest energy from the water flow. When in port shore power could be used to top the charge up. Flexible solar panels can be fitted to the sails, again giving some change to the batteries. The system is similar to that proposed by the Greenheart project [4]. As well as providing power storage the batteries will work as permanent ballast further aid the stability of the vessels.

There are various options available for the rig of the vessels. The principal requirements are that the rig is aerodynamically and economically efficient, that it can be reliably automatically controlled and that it has minimal impact on cargo handling operations. There is much work to be done choosing and optimising the rig which it is beyond the scope of this paper. A non-exhaustive list of options using extant developed technologies would include Omer wingsails, Aerorigs or Dynarigs, in both centre line and offset installations, Indo rig, biped rigs or spritsail rig..

By moving from a crewed vessel to an autonomous one the necessity to provide space and equipment for the crew is removed, meaning that for a given Gross Tonnage more hull volume is available as hold space. Also by moving to the use of renewables as the principal power source the volume taken by bunkers is reduced.
It is envisaged that the design of the autonomous control systems and support services will be shared with other projects.

3. AIMS AND OBJECTIVE

3.1 AIM OF THE PROJECT

This project aims to identify a suitable route, that may be used to test proof of concept vessels to demonstrate the commercial and functional viability of small autonomous wind powered cargo vessels. Over 60% of marine accidents can be attributed to human error[5].Once the vessels are fully developed and proven it is expected that they will be as reliable and safe as manned vessels and hence able to work in any waters. However here the aim is to choose areas where the operation will be challenging enough to test and improve the capabilities of the vessels without providing unnecessary obstacles to demonstrating their viability.. Whilst it will be beneficial for the final vessels to be as flexible as possible in terms of operating conditions and cargos, in order to develop an initial design some service parameters need be established to provide a framework within which to design the vessel. Choosing suitable ports and cargos for the trials at this stage will help define these design parameters. Once the principal design parameters are selected a better estimate of the build and running cost will be possible, enabling an outline business case to be made. It will also be helpful to have a concrete example of a potential use to help more easily communicate the concept in the future.

3.2 OBJECTIVES OF THE PROJECT

There are 5 interlinked objectives used to achieve the project aim.

- Areas with suitable physical and human geography are identified.
- Suitable ports within these areas are identified.
- Suitable cargos transported from near these ports are identified.
- Based on carrying these cargos an outline vessel specification is costed.
- The costs of carrying the cargo on the WASV are estimated, and compared with the likely costs of more conventional systems.

4. RESEARCH PROCEDURES

These five objectives require three distinct types of research to be undertaken.
The first section concentrates on the process of choosing a geographical area and selection of ports to focus the later work on.
The second stage looks at finding suitable cargos in the area to transport.
Finally an assessment of the costs associated with using the WASV to carry cargo on this route is made.
The intention in this study is not to complete an exhaustive search for suitable ports and cargos, but to find one or two possible applications on which to potentially test the viability of WASV for cargo transport.

4.1 AREAS AND PORTS

The purpose of this first section of the work is principally to narrow down the areas that need more thorough investigation in later stages of the project to manageable proportions. A list of criteria for suitable areas and ports were developed. Data from various online sources were combined and geocoded to highlight areas that were likely to have conditions that favoured the applications of the WASV. Ports in the selected areas were then assessed and marked qualitatively out of 3 on approach, entrance, and existing infrastructure and given a value of 0 or 1 depending on if they are currently commercially active.

These scores could then be combined to get a picture of the suitability of the ports in the area. The coast of the UK was worked clockwise from the Bristol Channel. Once an area with a sufficient density of likely ports was identified this stage was stopped and attention was moved onto searching for suitable cargos.

4.2 CARGO.

It is thought most likely that bulk cargos will form the most favourable goods for these vessels to first trade in. In order to minimise the potential pollution from an accident and for the reasons of public perception pointed out in the MUNIN short sea shipping report [6] oil and chemical tankers were avoided. Once an area containing suitable ports had been identified the principal local industries were identified and approached. Work was concentrated on situations where the current transport options were thought to be inadequate.

4.3 COSTING

Once a cargo and route had been chosen an initial assessment of the commercial feasibility could be made. An estimate of the build cost of the vessel was made by reference to vessels with similar characteristics at the building or planning stage and by reference to estimating techniques used in the shipbuilding industries. Some assumptions about the possible performance of the vessel were then made. This was used in conjunction with a sample of weather data to give the range of voyage times to be expected on the chosen route by using commercially available routing software. Cargo handling costs and port fees were taken from the publicly listed prices.
Costing the equivalent road trip was done using the Road Haulage Association cost tables.
Costing of the existing seaborne route was based on an estimate of the new build cost of a suitable coaster from which was calculated the minimum time charter rate. This was thought to offer the fairest comparison with the WASV costing as it was effectively comparing the cost of new infrastructure choices to serve a transport need. As a final check the costs of running the Timberlink service implicit in the subsidy regime were used to give upper and lower limits of what the conventional shipping costs may be.

5. ASSUMTIONS MADE

For the costings it has been assumed that sufficient other autonomous marine systems are in use that the supervision centres envisaged in the MUNIN report [7] are active and able to offer supervision services to WASV. Current practise in the use of ASV is that every vessel has a dedicated supervisor, and at present the situational awareness of commercially available surface vessels makes this necessary as it is not sufficiently advanced to avoid small obstacles. Work is ongoing to improve these capabilities and this project assumes that this work is successful.

Since no vessel design for the WASV exist (and part of the purpose of this project was to define the parameters that will allow that design to be created) an approximation of the polar for the vessels was chosen for initial estimates the voyage times.

Currently under the colregs uncrewed vessels would be of doubtful legality. There are however moves to address this issue with the Marine Autonomous Systems Regulatory Working Group (MASRWG), a body made up with representation from industry, academia and regulatory authorities such as the MCA, looking at how the legal status of autonomous vessels of all sorts can be managed. James Fanshawe, the chair of MASRWG has been contacted and made aware of this project. Although regulatory change is a slow process steps are being taken to start the process and it is expected that by the time these vessels are ready for deployment the colregs will have been changed to accommodate autonomous vessels. For this work it has been assumed that these vessels will be legal and insurable.

6. RESULTS AND DISCUSSION

6.1 THE CHOICE OF AREA AND PORT.

The waters around the coast of Great Britain provide some surprisingly challenging waters to test a vessel in, with areas with strong tides and large range of weather conditions. There are significant numbers of other vessels using the area with large volumes of cargo, fishing, naval and leisure traffic in some areas. Working on charts of the UK coast and other mapping sources areas were assessed based on the following criteria:

6.1(a) Traffic Density.

Data available from marinetraffic.com gives an indication of traffic density based on AIS signals. This was used to count out areas of high traffic density. It also highlighted areas with offshore installations such as wind farms. Although AIS reporting doesn't give a full report of traffic density as it only records the position of vessels fitted with transmitters it gives an adequate estimate to avoid areas with high traffic levels.
6.1(b)Traffic Separation Schemes (TSS)

Areas where TSS's operate were avoided, partly as they are an indication of a high traffic density, but also as the rules for their use limit the freedom for a sailing vessel to choose an optimum course based on the prevailing weather conditions.

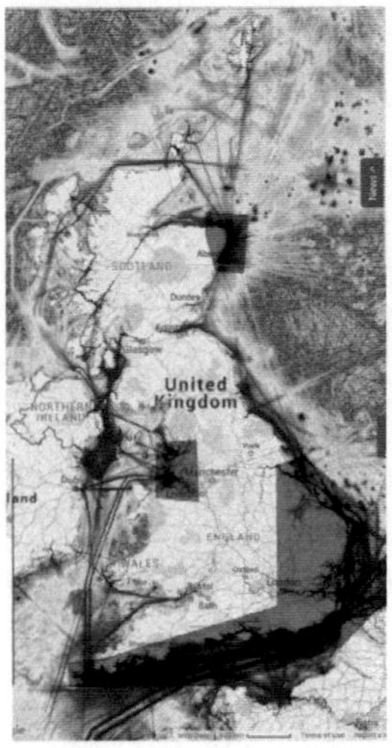

Figure 1 Areas excluded through traffic density.

6.1(c)Topographic Features.

Sailing vessels of the size envisaged in this study are quite maneuverable, so able to safely navigate through relatively constrained waters, however to avoid unnecessary problems in the early trials narrow passages were avoided. Areas where a direct sea route gives an obvious distance saving when compared with the direct land route were favoured.

Once areas with excessive traffic, TSS's and the like had been discounted it still left a significant proportion of the coastline available for investigation, with too many ports to be assessed in a timely manner. Areas of the coast where there is a shorter direct line by sea than by land (from North Cornwall to South Wales for example) would on first analysis be the most likely to offer a cost benefit over existing land routes and so would offer an incentive to choose an area. A set of areas with favorable geography was highlighted where road transport faced a considerably longer, or in some cases non existent, route than a sea based alternative. Clockwise from the Scilly Isles 8 potential areas were identified as having geographically attractive features, and a reasonable port density. The Bristol Channel, Cardigan Bay, The Northern Irish Sea, The Solway Firth, The Firth of Clyde, The Hebrides, The Moray Firth and the Firth of Forth.

Area	Comments
Bristol Channel	Insufficient active ports identified on south side.
Cardigan bay	Insufficient active ports identified and unlikely to be sufficient mileage saving.
Solway Firth	Insufficient active ports identified on north side.
The Northern Irish Sea	Some suitable ports on both sides. One potential cargo initially identified.
The Firth of Clyde	Very many suitable active ports, many smaller inactive ports that may be returned to use. Suitable cargos, poor road infrastructure.
The Hebrides	Very many suitable ports, many smaller inactive ports that may be returned to use, suitable transport problems, suitable cargos.
The Moray Firth	Insufficient active ports identified on north side.
Firth of Forth	Many Ports, several commercially active, first investigations indicate that scale of active ports may be unsuitable for WASV early trials.

Table 1 Geographically suitable areas investigated

6.1(d) Ports.

The application of the area criteria above reduced the number of ports to be investigated. Although one of the long term objectives of this project is to return unused port infrastructure to work, for initial trials is it thought appropriate to choose facilities that are currently in use. Port availability fed back into area choice, as lack of suitable active infrastructure eliminated some areas. Of the eight potential areas identified five were quite rapidly discounted. A summary of the logic behind this is shown in table 1 below.

Initially ports in areas thought likely to be favourable were assessed for suitability against the criteria described below by the use of available charts, pilotage notes, and port information freely available on various websites.

A 25m sailing vessel is a relatively handy size that is able to sail into quite constrained waters, however it is not wise to intentionally put a vessel into a situation where in order to avoid incident manoeuvres need to be completed flawlessly and that there is no space to, for example, recover from being caught in irons after a failed tack Thanks to the good quality of cartography in the charting of UK waters isolated hazards on an approach would not serve to discount a port from consideration (if the vessel is not able to reliably miss a charted, isolated hazard the navigation system would not be fit for purpose). The more difficult the port entrance the more likely it is that auxiliary power will be needed to successfully, and repeatedly, navigate the entrance. In order to favour wind power as much as possible (and hence lower transport emissions as much as possible) easy entrances were preferred. For this reason harbours with a long narrow approach were discounted. This was left as a subjective judgment of suitability rather than stating quantitative measurements.

Ports with locked entrances were discounted as they would double the number of docking procedures that would need to be undertaken on each voyage, so for this early project it was thought best to avoid this complication. An initial investigation of the areas showed that the Firth of Clyde and the Hebrides had a good combination of challenging road links, many small active ports and suitable cargos so work was concentrated on these areas.

6.2 CARGOS

Two principle classes of cargo were identified as being likely candidates for initial trials, products of quarries, particularly Glensanda quarry on the west coast of Loch Linnhe and timber, particularly from Kintyre. Contact with Aggregate Industries, Glensanda's operators discounted this as an initial option. Although they liked the concept, their business model for this operation depended on a larger throughput than could realistically be aimed for in early trials. As an alternative shipments of cement from Belfast to other depots round the coast were suggested. The quantity of material shipped on this route, and one of the destinations (Glasgow) initially made this an attractive route to develop. However on further investigation it was established that they moved no material between Belfast and Glasgow, and that the principle two destinations from Belfast were Ellesmere Port and Plymouth. Unfortunately these two destinations were in areas that had been discounted though traffic density (Ellsmere Port) and the presence of a nearby TSS (Plymouth). This has currently been taken no further, however it remains a potentially suitable cargo that should be investigated further as confidence in the WASV concept grows.

Timber transport from Kintyre proved to be a more immediately suitable possibility. Over 100000 tonnes of timber are moved out of the peninsula per annum[8]. The road route to the processing plants, mainly in Ayrshire, is long, busy, narrow and not suited to large lorries. This is already widely recognised, to the extent that in 2000 an alternative route using shipping was established. The Timberlink service transports timber from four ports (Campbeltown, Ardrishaig, Sandbank and Portavadie) to Troon where it is then moved by truck onto its final destination. When transported by road the logs are loaded from the in forest transport straight onto the trucks that move them to their final destination. When transported by sea the logs are moved from in forest transport to truck to ship to truck to final destination. The length of the route is shorter than would usually found viable for sea transport to be used, the vessel spends significantly more time being loaded or unloaded than in transit. These two extra modal shifts combined with the short route mean that even though marine transport is usually cheaper than road the marine route needs a subsidy to be commercially competitive with the road route. The subsidy is justified on the grounds of the various environmental and social benefits that it brings, less road damage, reduced impact on communities on the road route and reduced CO_2 emissions. These benefits are presented in a report [9] and give a useful baseline with which to compare the environmental impacts of a WASV based system.

6.3 COSTS

To cost the use of the WASV an approximation of their build cost is necessary. Lacking a design there is very little information to calculate a build cost from. Published figures for one vessel recently built and one at the planning stage were found that are sufficiently similar to the WASV concept in size and build style were found. These were combined with figures obtained from SPAR associates to give a range of possible build costs. It has been assumed that series production of the vessels will give a 30% reduction in their costs compared to one off production. Future more accurate costings will rely on more accurate hull and rig designs.

	One off	Series
Thames barge for Seachange Trust	£600000	£420000
Greenheart	£460000	£322000
WASV range used	£700000	£500000

Table 2 Build cost estimates 150dwt WASV

Costs for the remote supervision of the vessels have been taken from the MUNIN report on the autonomous vessels[10] The Munin costs are for the supervision of far larger vessels and assume that one operator can supervise 6 vessels.

Approximate costs have been worked out for moving timber with WASV using a variety of assumptions about both the building and operating costs and the costs and organisation of the logistics at each end. As can be seen in figure 2 unusually, for shipping, the costs of operating this service would not be mainly driven by the capital costs of the vessels, as the port costs represent a far more significant proportion of the total. Simulating the sailing legs from all four ports served by TimberLINK to Troon it appears that it is likely that a WASV could reliably move one load per day between any of the ports and Troon, allowing loading, voyage and unloading to be completed within the working days of the ports, with the return leg being undertaken overnight

For simplicity this report concentrates on three scenarios.

- A baseline case as a direct replacement for Timberlink.
- A case where port costs at both ends are equal and at the lower of the two current costs.
- A case where the cargo is loaded directly from the harvest site

In the first case where the WASV are operated as a straight replacement for the Timberlink service it is apparent that they could operate at a cost close to that of the straightforward road route (Figure 3). The service would maintain the environmental benefits identified for the Timberlink service. In addition it would reduce CO2 emissions by about 2kg/tonne in co service would maintain the environmental benefits identified for the Timberlink service. In addition it would reduce CO2 emissions by about 2kg/tonne in comparison to the conventional ship by principally using wind power to move across the Clyde. Also by shipping timber away 150 tonnes at a time, rather than in 1100 tonne consignments, the in port traffic issues identified in the environmental impacts report should be reduced.

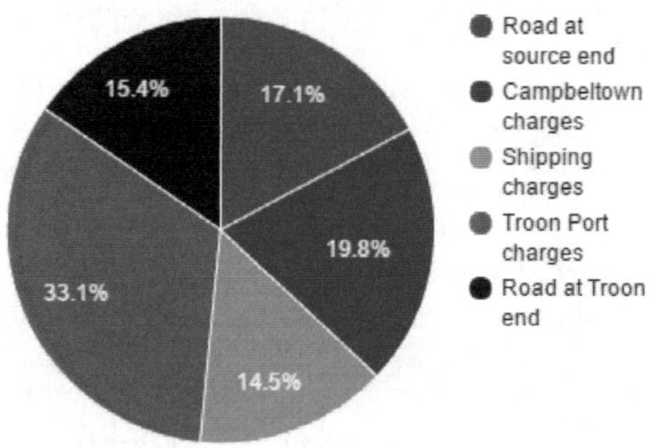

Figure 2 Cost contribution from different trip stages.

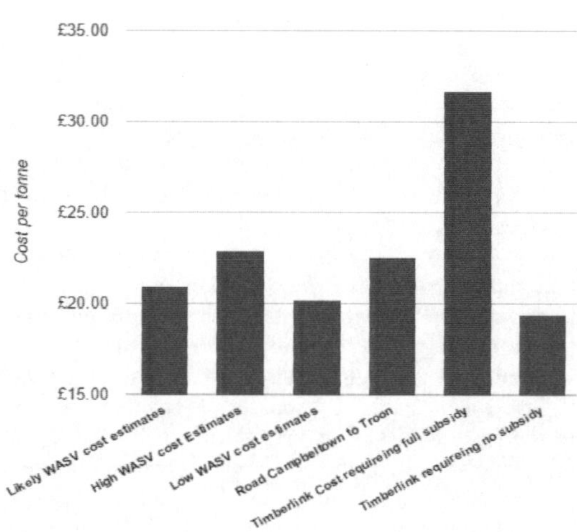

Figure 3 WASV cost £/Tonne compared with other transport options.

For the second case part of the rationale behind this project was to bring back into use under used port infrastructure. The hope is that doing this would increase price competition between ports, particularly as being smaller vessels the port infrastructure costs associated with using the WASV should be smaller than that associated with a conventional ship. To represent this scenario a set of operating costs have been calculated assuming that the port costs at the Troon end are the same as the much lower costs at the Campbeltown end. This demonstrates how sensitive to port costs trips of this length are.

Finally for the third case a set of costs have been calculated for using the vessels working from a temporary jetty such as the ones setup by JST Ltd. These would facilitate the transfer of the timber straight from the forest to the WASV, so removing the costs and environmental impacts associated with the transfer by road from the wood to the port (Figure 4). By eliminating another part of the road transport, and another loading procedure CO_2 emissions are reduced by a further 3kg/tonne. The small size of the WASV means that the number of vessels operating to a particular place can be matched to the production rate off the site meaning that timber could be loaded at approximately the rate it is produced.

.

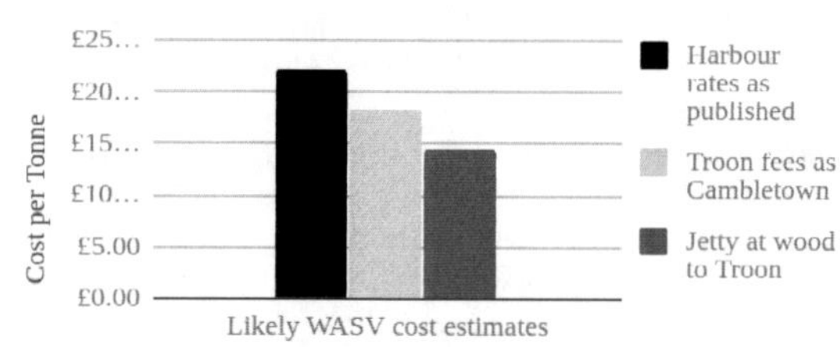

Figure 4 Comparison of different operating procedures

As was pointed out earlier the appropriate size of vessel for a particular route is a function of the loading time and the voyage time. In general as voyage time reduces so does optimum vessel size. To check that the 150dwt vessels are an appropriate size to start with a comparison was worked out assuming a 75dwt vessel and a 300dwt vessel (Figure 5). The assumption has been made that over this range of sizes the loading rate will stay constant, as at 300dwt there is unlikely to be sufficient length for two cranes to work simultaneously. The calculation shows that the 300dwt vessel is probably slightly cheaper per tonne than the 150dwt however the longer loading time needed for the larger vessel means the probable round trip time is sufficiently slowed so the utilization of the vessel would decrease as it could no longer be reliably loaded and unloaded on a daily basis. On these grounds 150 dwt does appear to be a suitable cargo capacity to aim for in the initial design stages.

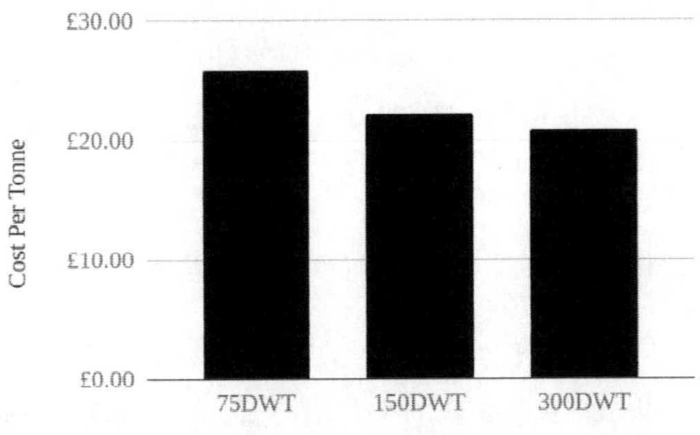

Figure 5 Comparison of transport cost against vessel size.

Once an initial design has been made a more accurate set of build costs and polars will be available, enabling a more accurate round of optimisation to be undertaken. The issue of how the round trip time interacts with ensuring the vessel arrives at its destination during port working hours is a factor that has only been approximately modelled in this study. It is however clear that with a short voyage time the effect of a delay to wait for operating hours would represent a larger proportion of voyage cost than it would for a longer voyage. The use of several smaller WASV to replace a single larger conventional coaster should enable the cargo capacity deployed on a route to more easily be matched to the requirements.

Although the route across the Clyde is the only one that has been modelled trying to look at all costs involved in the transport a comparison of the voyage costs has been made for a variety of voyage lengths, assuming that the WASV travels at ½ the speed of a conventional ship (Figure 6). As the trip gets longer the larger vessel gets cheaper. This however assumes that both sizes of vessel can be employed equally efficiently. If the consignment of goods is smaller than the payload of the vessel the scale related efficiency gains are negated.

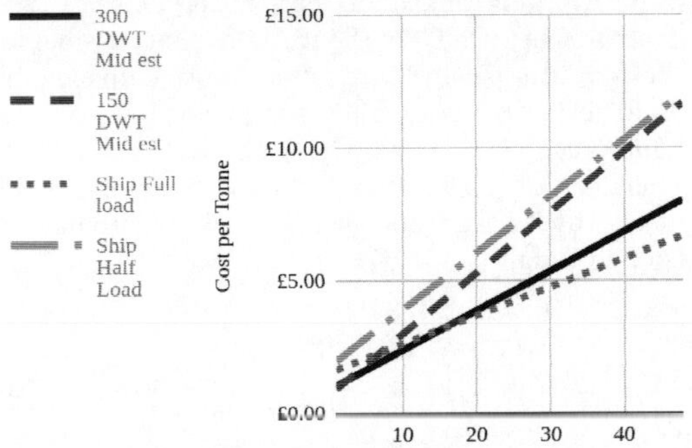

Figure 6 Comparison of shipping only costs (loading and sailing time, but no port fees) 150 and 300DWT WASV against fully laden and ½ laden 1100 DWT coaster.

Contact with potential clients for both the cement and timber industries raised the issue of who would be responsible for stowing and securing the cargo on the vessels in the absence of a crew. This is an issue that would need addressing.

Both the cargos identified so far in this research as candidates for carriage, cement and timber are members of the minor bulk trades. They often need specialised storage or handling equipment, reducing the available fleet size that is able to compete for their carriage. They are frequently moved between fixed points with vessels on long time charter agreements or on COAs. This area may be the most appropriate to look to for the further development of WASV.

7. STUDY LIMITATIONS

Although one of the long term aims of this project is to exploit currently under used port infrastructure for the first trials ports that had at least a certain degree of commercial activity were sought so the vessels could be brought into use with the minimum of additional investment. The final search for cargos and the costing of routes was only carried out on only one of the areas identified as potentially suitable. Further investigation of the other areas noted as being particularly suitable may well yield other potential uses for the WASV.

The very general nature of the outline design of the WASV at the moment means that costing data for their use can only be approximate. This study solely looked to establish if there is a reasonable likelihood that there would be viable applications for the vessels if developed, and costings were done on the assumption that there is a mature autonomous shipping industry. It in no way addresses their likely development costs.

Although some contacts were made with arguably the most important stakeholder group in the future development the WASV, the future customers for their transport services, the nature of this project was that they were only identified towards the end so contact was relatively limited.

8 CONCLUSIONS

This project has successfully identified that there are some ports and cargos that the WASV concept may be suitable to serve, and that further development of these vessels is worthwhile. Three areas of work necessary to develop these vessels have been identified

- Completing an outline design of the vessel
- Developing the rig and its control systems
- Development of the regulatory and management structure to safely operate the vessels

In this study timber movement on the west coast of Scotland has been identified as a good cargo for the WASV to be developed for, and the minor bulk trades in general have been noted as an area worthy of further investigation. To check the initial findings of this work a vessel design should be progressed to the point where a more reliable estimate of the build cost is possible, and a VPP can be run to produce an accurate set of polars for the vessel. The performance of the vessels should then be more fully simulated using weather records from several years, rather than the initial fortnight long runs undertaken here. As well as the raw voyage times the way the trips integrate with port working hours should be confirmed and contacts with potential future customers should be followed up.

Development of the rig and its management software is strongly related to the core business of OneSails (design and manufacture of sails). In order for the VPP to be reliable the design of the rig to be used will need to be finalized. As well as the use on WASV this technology may also have direct applications to our current markets and be transferable to other projects looking at the wider use of wind power in shipping. This work would also give us the opportunity to involve spar makers, riggers, deck hardware suppliers and software specialists. It will require investment in physical hardware for prototyping and testing. To progress this work stream potential industry partners need to be identified, brought together and a budget found.

The work needed on the management and operation of the vessels, covering mundane but essential aspects such as who is responsible for the safe stowage of the cargo, how the mooring of the vessels is undertaken and aspects of the autonomous system design that are more central to the vessels safe operation, such as the avoidance of small vessels without AIS, is common to all the potential applications of autonomy to cargo vessels. this has already been identified as a priority by the UKMIA. Contact with the MASRWG has alerted them to the need to include wind powered vessels in their work on updating the COLREGS to include autonomous vessels.

9. ACKNOWLEDGEMENTS

This work was carried out with TTRIG funding from the DfT.

10. REFERENCES

1. CORRES A 'Open issues concerning the future of the European Shortsea Fleet.' *Shortsea Shipping Days. 2019.*

2 INSTITUTE FOR MANUFACTURING (IfM), (2015) "UK Marine Industries Technology Roadmap 2015" *Retrieved 13/9/19 from* https://www.maritimeindustries.org

3 RODSETH O (21/09/2015) 'Impact on Short Sea Shipping' *retrieved 15/8/2019 from www.unmanned-ship.org*

4 GREENHEART PROJECT (2015) Auxiliary Power System *Retrieved 1/3/2017 from http://greenheartproject.org*

5 Wilko C. Bruhn et al (2014) "Can unmanned ships improve navigational safety?"*Transport Research arena 5th Conference*

6 RODSETH O (21/09/2015) 'Impact on Short Sea Shipping' *retrieved 15/8/2019 from www.unmanned-ship.org*

7 KRETSCHMANN L "Maritime Unmanned Navigation through Intelligence in Networks D9.3: Quantitative assessment" (2015) *retrieved 15/8/2019 from www.unmanned-ship.org*

8 TRANSPORT and TRAVEL RESEARCH Ltd 2010 "TimberLINK Environmental Benefits Review" *retrieved 1/3/2017 from* http://scotland.forestry.gov.ukf

9 TRANSPORT and TRAVEL RESEARCH Ltd 2010 "TimberLINK Environmental Benefits Review" *retrieved 1/3/2017 from* http://scotland.forestry.gov.ukf

10 7 KRETSCHMANN L "Maritime Unmanned Navigation through Intelligence in Networks D9.3: Quantitative assessment" (2015) *retrieved 15/8/2019 from* www.unmanned-ship.org

11. AUTHORS BIOGRAPHY

Alistair Chaplin holds the current porition of Sail Designer at OneSails GBR (South). He is responsible for the design and manufacture of all sails delivered by OneSails to the British market, and for the development of our place in the newly emerging market for commercial sail. This work grows from the work originally undertaken for his MSc in environmental decision making.

The Royal Institution of Naval Architects
8-9 Northumberland Street
London, WC2N 5DA
United Kingdom

ISBN 978-1-7138-0872-5